Bioinformatics and Genomes

Current Perspectives

Edited by

Miguel A. Andrade

EMBL, Heidelberg, Germany

Horizon Scientific Press
P.O. Box 1
Wymondham
Norfolk NR18 0EH
England

www.horizonpress.com

British Library Cataloguing-in-Publication Data

A catalogue record for this book is available from the British Library

ISBN:1-898486-47-6

Description or mention of instrumentation, software, or other products in this book does not imply endorsement by the author or publisher. The author and publisher do not assume responsibility for the validity of any products or procedures mentioned or described in this book or for the consequences of their use.

Printed and bound in Great Britain
by Antony Rowe Ltd, Chippenham, Wiltshire

Contents

Books of Related Interest

Genomes and Databases on the Internet	2002
Emerging Strategies in the Fight Against Meningitis	2002
Microbial Multidrug Efflux	2002
Environmental Molecular Microbiology: Protocols and Applications	2001
The Spirochetes: Molecular and Cellular Biology	2001
Development of Novel Antimicrobial Agents	2001
H. Pylori: **Molecular and Cellular Biology**	2001
Flow Cytometry for Research Scientists: Principles and Applications	2001
Prokaryotic Nitrogen Fixation: A Model System for the Analysis of a Biological Process	2000
NMR in Microbiology: Theory and Applications	2000
Oral Bacterial Ecology: The Molecular Basis	2000
Molecular Marine Microbiology	2000
Cold Shock Response and Adaptation	2000
Prions: Molecular and Cellular Biology	1999
Probiotics: A Critical Review	1999
Peptide Nucleic Acids: Protocols and Applications	1999
Intracellular Ribozyme Applications:Protocols and Applications	1999

Contributors

Patrick Aloy
European Molecular Biology Laboratory
Meyerhofstrasse 1
69117 Heidelberg
Germany
Email: aloy@embl-heidelberg.de

Miguel A. Andrade
European Molecular Biology Laboratory
Meyerhofstrasse 1
69117 Heidelberg
Germany
Email: andrade@embl-heidelberg.de

Chantal Abergel
Information Génétique et Structurale
UMR 1889
31 Chemin Joseph Aiguier
13 006 Marseille
France
Email: chantal.abergel@igs.cnrs-mrs.fr

Peer Bork
European Molecular Biology Laboratory
Meyerhofstrasse 1
Heidelberg 69117
Germany
Email: bork@embl-heidelberg.de

Birgit Eisenhaber
Research Institute Molecular Pathology
Dr. Bohr-Gasse 7
A-1030 Vienna
Republic Austria
Email: Birgit.Eisenhaber@imp.univie.ac.at

Frank Eisenhaber
Research Institute Molecular Pathology
Dr. Bohr-Gasse 7
A-1030 Vienna
Republic Austria
Email: Frank.Eisenhaber@imp.univie.ac.at

Richard A. George
Division of Mathematical Biology
National Institute for Medical Research
The Ridgeway
Mill Hill
London NW7 1AA
U.K.
Email: rgeorge@nimr.mrc.ac.uk

Jaap Heringa
Division of Mathematical Biology
National Institute for Medical Research
The Ridgeway
Mill Hill
London NW7 1AA
U.K.
Email: jhering@nimr.mrc.ac.uk

Jens Kleinjung
Division of Mathematical Biology
National Institute for Medical Research
The Ridgeway
Mill Hill
London NW7 1AA
U.K.
Email: jkleinj@nimr.mrc.ac.uk

Sebastian Maurer-Stroh
Research Institute Molecular Pathology
Dr. Bohr-Gasse 7
A-1030 Vienna
Republic Austria
Email: stroh@imp.univie.ac.at

Enrique Merino
Instituto de Biotecnología
Univ. Nacional Autónoma de México
AP 510-3
Cuernavaca Mor
62250 México
Email: merino@ibt.unam.mx

Eric Minch
LION Bioscience AG
98 Waldhoferstrasse
69123 Heidelberg
Germany
Email: minch@lion-ag.de

Enrique Morett
Instituto de Biotecnología
Univ. Nacional Autónoma de México
AP 510-3
Cuernavaca Mor
62250 Mexico
Email: emorett@ibt.unam.mx

Cédric Notredame
Information Génétique et Structurale
UMR 1889
31 Chemin Joseph Aiguier
13 006 Marseille
France
Email: cedric.notredame@igs.cnrs-mrs.fr

Seán I. O'Donoghue
Lion Bioscience AG
Waldhoferstr. 98
69123 Heidelberg
Germany
Email: sean.odonoghue@lionbioscience.com

Baldomero Oliva
IMIM / UPF
c/ Doctor Aiguader 80
08003 Barcelona
Spain
Email: boliva@imim.es

Leticia Olvera
Instituto de Biotecnología
Univ. Nacional Autónoma de México
AP 510-3
Cuernavaca Mor
62250 México
Email: lolvera@ibt.unam.mx

Maricela Olvera
Instituto de Biotecnología
Univ. Nacional Autónoma de México
AP 510-3
Cuernavaca Mor
62250 México
Email: mari@ibt.unam.mx

Carolina Perez-Iratxeta
European Molecular Biology Laboratory
Meyerhofstr. 1
69117 Heidelberg
Germany
Email: cperez@embl-heidelberg.de

Emmanuvel Rajan
Instituto de Biotecnología
Univ. Nacional Autónoma de México
AP 510-3
Cuernavaca Mor
62250 México
Email: emanuvel@ibt.unam.mx

Robert B. Russell
European Molecular Biology Laboratory
Meyerhofstrasse 1
69117 Heidelberg
Germany
Email: russell@embl-heidelberg.de

Gloria Saab-Rincon
Instituto de Biotecnología
Univ. Nacional Autónoma de México
AP 510-3
Cuernavaca Mor
62250 México
Email: gsaab@ibt.unam.mx

Mikita Suyama
European Molecular Biology Laboratory
Meyerhofstrasse 1
69117 Heidelberg
Germany
Email: suyama@embl-heidelberg.de

Javier Tamames
ALMA bioinformática
Centro Empresarial Euronova
Ronda de Poniente
28760 Tres Cantos
Madrid
Spain
Email: tamames@almabioinfo.com

David Torrents
European Molecular Biology Laboratory
Meyerhofstrasse 1
69117 Heidelberg
Germany
Email: torrents@embl-heidelberg.de

Preface

Picture yourself as a clock master, a specialist in repairing old mechanical clocks. Your work is arduous but you love it! You have accumulated skill over the years and you have an expert way of working.

Old clocks are built in different ways and from time to time you face a clock that is a real problem: you cannot figure out how the mechanism works. However, using your intuition, and by trial and error, you usually discover how to fix the clock. For example, you first open the clock and then you find out how some of the pieces of the clock are connected. You guess that one piece is missing or broken. You try to replace that piece and then you see whether the clock works again or not. You turn some knob somewhere and then the clock runs faster. In summary, you might not understand how the whole clock works, but you have a rough idea of the important parts and as a rule you find your way and you eventually succeed in repairing the clock.

And then, one morning, there is this little funny guy with thick glasses who comes into your workshop and stands nervously by the counter. He tells you he wants to work in your shop as an apprentice. 'I could hardly afford to pay you. These are bad times.' you say. 'But I see that you have a lot of work to do!' he insists pointing at the shelves where clocks waiting to be repaired accumulate. 'Yes. I didn't say the contrary. I just need a lot of time to repair each clock and then the wage is not so good.' you reply. 'What if I made your work faster?' the little guy answers. 'I can show you a new way of working with the clocks.'

You look sternly upon him but then decide to give him a chance. 'Show me what you are able to do. There you have that clock. What can you tell me about it?' you say. And you point him a costly clock from the 18th century from one of your shelves. Surely it's a tough job.

And the guy grabs the clock and then your toolbox. He is very skilled with the tools. Obviously, he has some practice and good fingers. You stare surprised at his elegant movements. But, too late, you realise with dismay that he has completely taken apart the expensive clock, and now he is very carefully laying over your table all the pieces side by side.

He is even classifying the objects: there are seven springs, twelve small wheels with teeth, the dials, two big wheels, a metal ring, a circular glass plate, etc. and many more pieces that you cannot even work out what they are. As you examine the pieces (some of which you already know but some not) you may even realise that a few have been broken in the process of taking them apart. Or that a piece is really made of other smaller pieces. In fact, you are less sure now of whether you are looking at the real pieces that make up the clock.

'Now you know everything about the clock! Now you can repair it!' exclaims the guy who wanted to help you with a smile. You shake your head. 'I think I have the pieces,' you reply warily; 'but, I cannot tell you what this is for', and you pick at random one little twisted metal thingy. 'But you can check that piece, can you not?' the little man insists. You doubt. 'Well… I don't even know where this was in the original clock, let alone to which pieces it did connect. I might have to take another similar clock and search for the piece inside. That may take a lot of time but now I cannot decide whether this piece is important or not for the clock to work.' The little man has a new idea: 'I will take apart another machine and compare the pieces.' And he leaves the shop before you can ask him whether he can assemble the dismantled clock again.

The next morning the little man comes back again with a triumphant face. You feel like strangling him about the wrecked clock but while reaching for his throat he produces a metal piece out of his pocket. 'Look what I found! I took apart my toaster and there inside was this piece that is similar to the one I found yesterday in your clock. It must be important!'

You look at the piece from the toaster. You compare that to the one from the 18th century clock and you are not sure. They look similar but, how can you know that they are really doing the same? How will that help you to fix the clock? And, after all, what does a toaster and a 300 year-old clock have to do with each other? The man tells you some gibberish about some P-values that demonstrate that both pieces must work the same but you don't understand, and you don't care anyway.

The little man wants to try something else. Well, he is really trying hard! He grabs several parts of the broken clock: 'Look! I think that all these six pieces come together and I think they have to work like this'. Now that's something. He has put together some of the wheels, and springs. You know that that almost makes sense. And the strange piece is in the middle so it must be important after all.

You have mixed feelings now. The small man is so helpful and sympathetic. You think that maybe you might really learn something from his strange ways. You don't see clearly how, but it would really be exciting to understand everything about every clock! 'All right! You are hired for a trial period,' you say; 'but I will not pay you much.'

'Oh! Thank you! Thank you ever so much! You will not regret!' The little guy is very proud and tells you he cannot wait to try his theories. He is unaware that you are hiring him not because you think that he is brilliant, but just to prevent him from going to the competition.

Akin to the moral of the above story, this book depicts a similar clash of styles. Its chapters will give you examples of how traditional ways of doing molecular biology are being confronted with the massive production of genomic data and with its subsequent computational analysis.

Since the British scientist Fred Sanger developed the first sequencing techniques in 1977, the sequencing of complete micro-organisms has now become routine, several eukaryotic genomes (including the first human versions) have been sequenced, the production of genomic data has exploded in quantity and the sequencing price per base has dropped beyond any previous expectation. Sequencing is cheap. But why do it?

Complete genomes promise quick solutions to fighting pathogens, solving human diseases, modifying organisms for our benefit, etc. at a much quicker pace than the laborious traditional molecular biology approach. At this point, it is not clear how much truth is in those promises. However, there are the first signs indicating that it might be so.

This book will give you some of those signs from the feather of bioinformaticians actively working on a variety of approaches to the analysis of genomic data. The expertise of these people ranges from detecting genes within genomes, gene expression, protein sequence alignment, protein domain analysis, protein structure/function analysis and prediction, post-translational modification, analysing and modelling metabolism, to aspects of genomic information organization, such as data base development and data visualization. As you will see, all of these aspects strongly inter-relate, making the computational analysis rich and interesting.

The last chapter of this book (which gathers the opinions of the authors of each of the chapters) tries to look ahead into the future of *Bioinformatics and Genomes* beyond the *Current Perspectives*, undoubtedly a difficult exercise in this quickly changing field. After all, who could have predicted the current directions that genomic analysis has taken say ten years ago?

A last note: Our visions are inevitably partial, and, as someone once said about invited reviews, they should be handled with care. But be sure that each chapter contains some provocative ideas that found a place here exactly because the book was collected with a much looser control than the one exhibited by often merciless journal peer reviewers.

If you are one of those clock-masters, may this book bring you closer to the new ways of the little funny man. If you are intending to become a little funny man/woman yourself, this book may give you some directions. In the meantime, let's wish we keep that job at the clock-shop beyond the trial period!

Miguel A. Andrade

From: *Bioinformatics and Genomes: Current Perspectives*
Edited by: Miguel A. Andrade

Chapter 1

Predicting Protein Structural Domain Boundaries From Sequence Data

Richard A. George, Jens Kleinjung
and Jaap Heringa

Abstract

Understanding the domain content of a protein and delineating the associated domain boundaries is a crucial step for many areas in protein science. Structural studies by NMR and X-ray crystallography are often greatly aided by such knowledge. Also, sequence database searches are typically enhanced when performed using sequence stretches corresponding to a single domain, because a domain is a recurring functional and evolutionary unit in proteins. In this chapter, we will review existing methods for predicting and delineating domain boundaries from sequence and structure information. We will also describe the available public databases with protein domain information. Finally, we will give an overview of our approaches for domain prediction. These include: attempts to predict linker fragments directly from observed differences between linker and non-linker segments; our method

SnapDRAGON, which exploits the consistency of domain boundary placement observed in a large set of ab initio 3D models generated for a given query sequence by a distance geometry-based method; and DOMAINATION, a method that predicts boundaries based on a new PSI-BLAST-related iterative protocol.

Biological Aspects Of Protein Domains

Domains can be defined as folded entities of protein structures formed by association of secondary structure elements. However, considering domains as structural building blocks implies to acknowledge them also as sequentially and functionally conserved elements in evolution. Protein families have emerged from common ancestors by different combinations and associations of domains (Bork, 1991; Doolittle, 1995; Heringa and Taylor, 1997). The size of individual structural domains varies widely from 36 residues in E-selectin to 692 residues in lipoxygenase-1 (Jones *et al.*, 1998), the majority (90%) having less than 200 residues (Siddiqui and Barton, 1995) with an average of about 100 residues (Islam *et al.*, 1995). Small domains (less than 40 residues) are often stabilized by metal ions or disulphide bonds. Large domains (greater than 300 residues) are likely to consist of multiple hydrophobic cores (Garel, 1992).

Domains are genetically mobile units, and multidomain families are found in all three kingdoms (Archaea, Bacteria and Eukarya) underlining the finding that ‘Nature is a tinkerer and not an inventor’ (Jacob, 1977). The majority of genomic proteins, 75% in unicellular organisms and more than 80% in metazoa, are multidomain proteins created as a result of gene duplication events (Apic *et al.*, 2001). Domains in multidomain structures are likely to have once existed as independent proteins, and many domains in eukaryotic multidomain proteins can be found as independent proteins in prokaryotes (Davidson *et al.*, 1993). For example, vertebrates have a multi-enzyme protein (GARs-AIRs-GARt) comprising the enzymes GAR synthetase (GARs), AIR synthetase (AIRs), and GAR transformylase (GARt)[1]. In insects, the polypeptide appears as GARs-(AIRs)2-GARt. However, GARs-AIRs is encoded separately from GARt in yeast, and in bacteria each domain is encoded separately (Henikoff *et al.*, 1997). Genetic mechanisms influencing the layout of multidomain proteins include gross rearrangements such as inversions, translocations, deletions and duplications, homologous recombination, and slippage of DNA polymerase during replication (Bork *et al.*, 1992).

[1] GAR: glycinamide ribonucleotide; AIR: aminoimidazole ribonucleotide

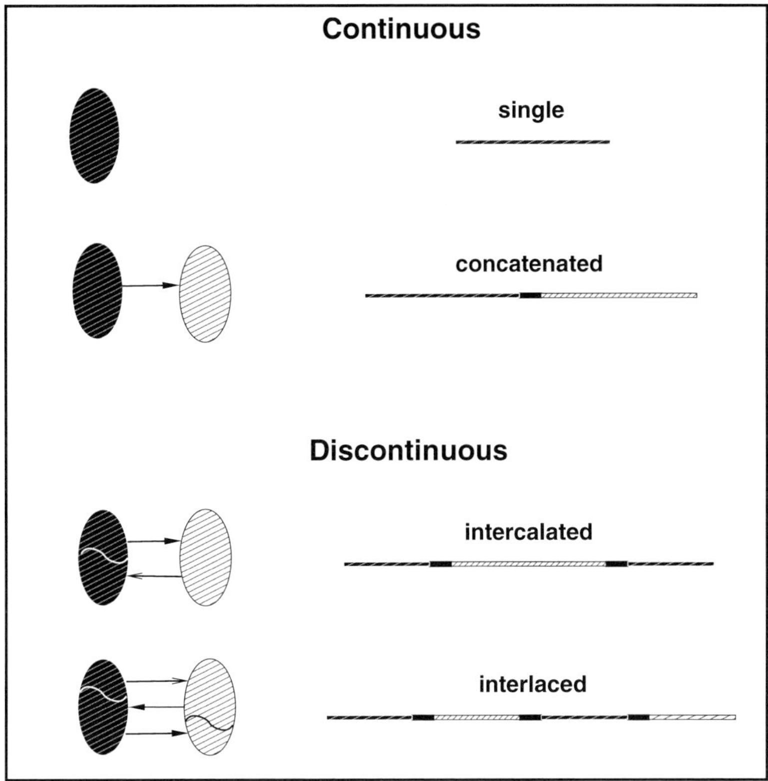

Figure 1. Schematic illustration of different types of connectivity in multidomain structures (left) and their sequences (right).

Although genetically conceivable, the transition from two single domain proteins to a multidomain protein requires that both domains fold correctly and that they accomplish to bury a fraction of the previously solvent-exposed surface area in a newly generated inter-domain surface. Domain swapping is such a structurally viable mechanism for forming oligomeric assemblies (Bennett *et al.*, 1995). In domain swapping, a secondary or tertiary element of a monomeric protein is replaced by the same element of another protein. Domain swapping can range from secondary structure elements to whole structural domains. It also represents a model of evolution for functional adaptation by oligomerization, e.g. of oligomeric enzymes that have their active site at sub-unit interfaces (Heringa and Taylor, 1997).

Evolution gives rise to various forms of domain arrangements that can be classified by their connectivity. Figure 1 displays the connectivity between domains, ordered by their degree of complexity (Das and Smith, 2000). The occurrence of interlaced proteins showing a high degree of connectivity is the main difference between a structural domain and a module. A module will be limited to one or two connections between domains, whereas structural domains can have unlimited connections within the given criterion of the existence of a common core. Thus, in some cases several structural domains can be assigned to a single module.

An understanding of the domain organisation of a protein sequence is important for many areas in protein science; examples can be found in NMR (Campbell and Downing, 1994; Pfuhl and Pastore, 1995; Castiglone-Morelli *et al.*, 1995), protein engineering (Guerois and Serrano, 2001), site directed mutagenesis experiments (Nielsen and Yamada, 2001) and the optimisation of structural prediction methods (Bonneau *et al.*, 2001). In the following sections we will describe domain databases, computational tools for delineating domains, and the impact of genomic data.

Domain Databases

Domain Sequence Databases

Since many proteins have multiple domains there is a need to characterise new sequences at the domain level. Many databases of protein motifs and domains have been developed for this purpose. Among these are BLOCKS, COGs, DOMO, Pfam, PRINTS, ProDom, PROSITE, PROT-FAM, SBASE, SMART and InterPro (see Table 1).

BLOCKS

Blocks are short multiply aligned ungapped segments corresponding to the most highly conserved regions of proteins. The rationale behind searching a database of blocks is that information from multiply aligned sequences is present in a concentrated form, reducing noise and increasing sensitivity to distant relationships. The BLOCKS Database (Henikoff *et al.*, 2000) is automatically generated by looking for the most highly conserved regions in groups of proteins documented in various domain databases. Version 13.0 of the BLOCKS database consists of 8656 sequence blocks generated specifically from proteins in the PROSITE database.

COGs

The COGs (Clusters of Orthologous Groups) database is a phylogenetic classification of the proteins encoded within complete genomes (Tatusov *et al.*, 2001). It primarily consists of bacterial and archaeal genomes. Incorporation of the larger genomes of multicellular eukaryotes into the COG

Table 1. Domain databases

Type	Database Name	URL	Reference
SEQUENCE	BLOCKS	http://blocks.fhcrc.org	Henikoff *et al.* (2000)
	COGS	http://www.ncbi.nlm.nih.gov/COG	Tatusov *et al.* (2001)
	DOMO	http://www.infobiogen.fr/services/domo	Gracy and Argos (1998b)
	InterPro	http://www.ebi.ac.uk/interpro	Apweiler *et al.* (2000)
	Pfam	http://www.sanger.ac.uk/Pfam	Bateman *et al.* (2000)
	PRINTS	http://www.bioinf.man.ac.uk/dbbrowser/PRINTS	Attwood *et al.* (1999)
	ProDom	http://www.toulouse.inra.fr/prodom.html	Corpet *et al.* (2000)
	PROSITE	http://www.expasy.ch/prosite	Hofmann *et al.* (1999)
	PROT-FAM	http://mips.gsf.de/proj/protfam	Srinivasarao *et al.* (1999)
	SBASE	http://www3.icgeb.trieste.it/~sbasesrv	Murvai *et al.* (2000)
	SMART	http://SMART.embl-heidelberg.de	Schultz *et al.* (2000)
STRUCTURE	3Dee	http://www.compbio.dundee.ac.uk/WWW_Servers/3Dee/3dee.html	Siddiqui *et al.* (2001)
	CATH	http://www.biochem.ucl.ac.uk/bsm/cath	Orengo *et al.* (1997)
	FSSP	http://www2.ebi.ac.uk/dali	Holm and Sander (1997)
	SCOP	http://scop.mrc-lmb.cam.ac.uk/scop	Murzin *et al.* (1995)

system is achieved by identifying eukaryotic proteins that fit into already existing COGs. Eukaryotic proteins that have orthologs within different COGs are split into their individual domains. The COGs database currently consists of 3166 COGs including 75,725 proteins from 44 genomes.

DOMO

The DOMO database contains domain multiple alignments automatically generated from successive sequence analysis steps including similarity search, domain delineation, multiple sequence alignment and motif construction (Gracy and Argos, 1998a). It has full coverage of the SWISSPROT and PIR sequence databases (Bairoch and Apweiler, 2000). The database currently holds 99,058 domains clustered into 8,877 multiple sequence alignments.

Pfam

Pfam is a collection of protein domain family alignments and profile-Hidden Markov Models (HMM) (Bateman *et al.*, 2000). Pfam is composed of two parts, PfamA and PfamB. PfamA is the curated section of Pfam and contains manually crafted multiple alignments and profile-HMMs for 3,071 domain families (version 6.6). PfamB covers all alignments contained in the ProDom database that are absent in PfamA; it comprises 57,477 domain families.

PRINTS

PRINTS is a database of protein fingerprints (Attwood *et al.*, 1999). A fingerprint is a group of conserved motifs used to characterise a protein family. Fingerprints can encode protein folds and functionalities more flexibly and powerfully than a single motif. Release 31.0 of PRINTS contains 1,550 entries, encoding 9,531 individual motifs.

ProDom

ProDom (Corpet *et al.*, 2000) is a database of protein domain families automatically generated from SWISSPROT and TrEMBL sequence databases (Bairoch and Apweiler, 2000) using a novel procedure based on recursive PSI-BLAST searches (Altschul *et al.*, 1997). Release 2001.2 of ProDom contains 283,772 domain families, 101,957 having at least 2 sequence members. ProDom-CG (Complete Genome) is a version of the ProDom database which holds genome-specific domain data.

PROSITE

PROSITE (Hofmann *et al.*, 1999) is a good source of high quality annotation for protein domain families. A PROSITE sequence family is represented as a pattern or profile. The profiles provide a means of sensitive detection of common protein domains in new protein sequences. PROSITE release 16.46 contains signatures specific for 1,098 protein families or domains. Each of these signatures comes with documentation providing background information on the structure and function of these proteins.

PROT-FAM

PROT-FAM (Srinivasarao *et al.*, 1999) is a curated database of homology clusters produced and maintained within the context of the PIR sequence database (George *et al.*, 1996). If detected homologies between members cover the entire sequences, clusters are named superfamilies. Otherwise, if similarities are only found to include local regions, clusters of so-called homology domains are created. PROT-FAM currently contains 8,538 superfamilies and 374 homology domains.

SBASE

SBASE domains are protein sequence segments with known structure and/or function (Murvai *et al.*, 2000). The boundaries of the domains are either previously defined within publications or determined by homology to domains with known boundaries such as given in the PROT-FAM and Pfam databases. The entries are clustered into 2,425 statistically validated domain groups (SBASE-A) and 739 non-validated groups (SBASE-B), release 9.0.

SMART

SMART (a Simple Modular Architecture Research Tool) contains profile-HMMs and alignments for each domain family (Schultz *et al.*, 2000). Alignments are based on known tertiary structures, where possible, or from homologues identified in a PSI-BLAST analysis (Altschul *et al.*, 1997). Alignments are checked manually for potential false positives or misassembled protein sequences derived from genomic sources. Database release 3.3 has 594 domain families found in signalling, extracellular and chromatin-associated proteins. The families are extensively annotated with respect to phyletic distributions, functional class, tertiary structures and functionally important residues. User interfaces to this database allow searches for proteins containing specific combinations of domains in defined taxa.

InterPro

Because the underlying construction and analysis methods of the above domain family databases are different, the databases inevitably have different diagnostic strengths and weaknesses. The InterPro database (Apweiler *et al.*, 2000) is a collaboration between many of the domain database curators. It aims to be a central resource reducing the amount of duplication between the databases. Release 3.2 of InterPro contains 3,939 entries, representing 1,009 domains, 2,850 families, 65 repeats and 15 post-translational modification sites. Entries are accompanied by regular expressions, profiles, fingerprints and Hidden Markov Models which facilitate sequence database searches.

Domain Structure Databases

Several methods of structural classification have been developed to classify the large number of protein folds present in the PDB. The most widely used and comprehensive databases are CATH, 3Dee, FSSP and SCOP, which use four unique methods to classify protein structures at the domain level.

A structural domain may be detected as a compact, globular substructure with more interactions within itself than with the rest of the structure (Janin and Wodak, 1983). Therefore, a structural domain can be determined by two shape characteristics: compactness and its extent of isolation (Tsai and Nussinov, 1997). Measures of local compactness in proteins have been used in many of the early methods of domain assignment (Rossmann *et al.*, 1974; Crippen, 1978; Rose, 1979; Go, 1978) and in several of the more recent methods (Holm and Sander, 1994; Islam *et al.*, 1995; Siddiqui and Barton, 1995; Zehfus, 1997; Taylor, 1999). However, this approach encounters problems when faced with discontinuous or highly associated domains and many definitions will require manual interpretation. Consequently there are discrepancies between assignments made by domain databases (Hadley and Jones, 1999).

CATH

The CATH domain database assigns domains based on a consensus approach using the three algorithms PUU (Holm and Sander, 1994), DETECTIVE (Swindells, 1995) and DOMAK (Siddiqui and Barton, 1995) as well as visual inspection (Jones *et al.*, 1998). The CATH database release 2.3 contains approximately 30,000 domains ordered into five major levels: Class; Architecture; Topology/fold; Homologous superfamily; Sequence family. Class is determined according to the secondary structure composition and packing within the structure. Three major classes are recognised; mainly-alpha, mainly-beta, and alpha-beta.

Architecture is the overall shape of a domain as defined by the packing of secondary structural elements, but ignoring their connectivity. The topology-level consists of structures with the same number, arrangement and connectivity of secondary structure based on structural superposition using SSAP structure comparison algorithm (Taylor and Orengo, 1989). A homologous superfamily contains proteins having high structural similarity and similar functions, which suggests that they have evolved from a common ancestor. Finally, the sequence family level consists of proteins with sequence identities greater than 35%, again suggesting a common ancestor.

CATH classifies domains into approximately 700 fold families; ten of these folds are highly populated and are referred to as 'super-folds'. Super-folds are defined as folds for which there are at least three structures without significant sequence similarity (Orengo *et al.*, 1994). The most populated is the α/β-barrel super-fold.

3Dee

The 3Dee structural domain repository (Siddiqui *et al.*, 2001) stores alternative domain definitions for the same protein and organises the domains into sequence and structural hierarchies. Most of the database creation and update processes are performed automatically using the DOMAK (Siddiqui and Barton, 1995) algorithm. However, some domains are manually assigned. It contains non-redundant sets of sequences and structures, multiple structure alignments for all domain families, secondary structure and fold name definitions. The current 3Dee release is now two years old and contains 18,896 structural domains.

FSSP

FSSP (Holm and Sander, 1997) is a complete comparison of all pairs of protein structures in the PDB. It is the basis for the Dali Domain Dictionary (Dietmann *et al.*, 2001), a numerical taxonomy of all known structures in the PDB. The taxonomy is derived automatically from measurements of structural, functional and sequence similarities. The database is split into four hierarchical levels corresponding to super-secondary structural motifs, the topology of globular domains, remote homologues (functional families) and sequence families.

The top level of the fold classification corresponds to secondary structure composition and super-secondary structural motifs. Domains are assigned by the PUU algorithm (Holm and Sander, 1994) and classified into one of five 'attractors', which can be characterised as all-α, all-β, α/β, α-β meander, and anti-parallel β-barrels. Domains which are not clearly assigned to a single attractor are grouped in a mixed class. In September 2000, the Dali classification contained 17,101 chains, 1,375 fold types and 3,724 domain sequence families. The database contains definitions of structurally conserved cores and a library of multiple alignments of distantly related protein families.

SCOP

The SCOP database (Structural Classification of Proteins) is a manual classification of protein structure (Murzin *et al.*, 1995). The classification is at the domain level for many proteins, but in general, a protein is only split into domains when there is a clear indication that the individual domains may have existed as independent proteins. Therefore, many of the domain definitions in SCOP will be different to those in the other structural domain databases. The principal levels of hierarchy are family, superfamily and fold, split into the traditional four domain classes, all-α, all-β, α+β and α/β. Release 1.55 of the SCOP database contains 13,220 PDB entries, 605 fold types and 31,474 domains.

Computational Tools For Domain Prediction

Delineating Domains by Comparative Sequence Analysis

DOMAINER

DOMAINER (Sonnhammer and Durbin, 1994) had previously been used to generate the ProDom database. The method begins with an all-against-all BLAST (Altschul *et al.*, 1997) comparison of all sequences in the SWISSPROT sequence database (Bairoch and Apweiler, 2000) to generate a list of homologous segment pairs that are clustered into homologous segment sets. Sequence positions between homologous segment sets are kept as links that indicate their relative location in a sequence, resulting in a large collection of graphs containing such links. Three situations indicate domain boundaries: real N- or C-termini, shuffled domains and repeating units. If a junction between two sets satisfies one of the above criteria, then the domain boundary is annotated and the corresponding graph is split. Finally, multiple alignments are constructed by concatenating the sets associated with each remaining graph.

MKDOM

The current release of ProDom uses MKDOM version 2 (Gouzy *et al.*, 1999), which is based on the assumption that the shortest sequence in a sequence database represents a single domain. This version of MKDOM iterates a loop of three steps. i) Initially the shortest sequence (>20 residues) is selected as a query from a non-redundant database; ii) PSI-BLAST (Altschul *et al.*, 1997) is run iteratively over the database until no more sequences are found or until a maximal number of iterations is reached, generating a domain family; iii) all domain sequences identified in the previous step are removed from the database, the remaining sub-sequences are returned to the database. The process is terminated when the database becomes depleted.

DOMO

The DOMO database (Gracy and Argos, 1998b) is constructed from an all-against-all comparison of residue composition of the sequences in the SWISSPROT (Bairoch and Apweiler, 2000) and PIR (George *et al.*, 1996) databases. Domain boundaries are inferred from the positional constraints derived from various detected pair-wise sequence similarities. The algorithm clusters overlapping pair-wise similarities into anchors. An anchor is defined as a set of similar protein sequences that align without gaps. Using the anchors, domain boundaries are assigned. These are inferred from true N- and C-termini or anchored repeats. The maximum length of a domain can be obtained easily if repeats within a single sequence are present. Following the thus identified boundaries, each sequence is spliced into all possible corresponding domain fragments. The resulting related sequence fragments for a domain family are multiply aligned by dynamic programming (Smith and Waterman, 1981). Each multiple alignment is extended toward the N- and C-termini, provided that acceptable levels of similarity are found.

PASS

The PASS (Prediction of Autonomous Folding Units based on Sequence Similarities) method uses a simple technique of domain delineation based on the stacking of sequences from a BLAST (Altschul *et al.*, 1990) search onto the query sequence (Kuroda *et al.*, 2000). Regions along a query sequence will often have a varying number of matching sequences in the BLAST results, leading to abrupt increases and decreases in sequence numbers along the query. Domain start and end positions are found by identifying such regions with a >20% change in the number of BLAST hits, provided the slope of the plateau following the change is <10% over 30 residues.

Multiple Domain Protein Diagnostic Patterns

Adams *et al.* (1996) described an algorithm that begins with a measure of sequence similarity between all pairs of sequences in a database using dynamic programming to generate a full similarity matrix (Adams *et al.*, 1996). A complete graph is generated in which each node corresponds to a sequence and the edges correspond to similarity scores between sequences. A connected graph with a minimum number of edges is constructed by dynamically filtering pair-wise similarities and deletion of the corresponding edges in the complete graph. The resulting graph is decomposed into "cliques", each clique representing a set of protein sequences potentially sharing at least one region of similarity above a threshold. Common element patterns are then constructed for those regions of similarity among the members of these maximal cliques (Smith and Smith, 1992) and sequences composing the clique of the highest cardinality are iteratively aligned via intermediate shared patterns. Each maximal clique pattern is subsequently matched to all sequences. Sub-sequences that match a pattern, and hence represent a domain, are removed and replaced by gap characters. The process is iterated until all regions containing a pattern common to at least two members have been identified and removed from the database.

GEANFAMMER

GEANFAMMER is a suite of programs that divides a set of protein sequences into families (Park and Teichmann, 1998). It has three main parts: a sequence matching procedure, clustering of the sequences, and domain cutting with DIVCLUS (Park and Teichmann, 1998). Initially, sequences are compared using an implementation of the Smith and Waterman algorithm (Smith and Waterman, 1981). Single linkage clustering is used to connect sequences to one another by similarity, using an upper E-value limit. DIVCLUS then inspects each cluster in an iterative manner and successively tests pairs of sequences for significance and extent of alignment between them. Three sequences are accepted as belonging to the same domain family if the matching regions overlap by more than 30 residues and the overlapping region represents over 70% of the overlap of the shorter of the two pairs. If three sequences are accepted as having a common homologous region, the common overlap segment becomes the new 'overlap sequence', which is tested against succeeding pairs of sequences. Accepted sequences are then merged. In this way non-matching segments are removed from a cluster.

GENERAGE

GENERAGE is an algorithm developed for the clustering of protein sequences between and within complete genomes (Enright and Ouzounis, 2000). BLAST (Altschul *et al.*, 1990) is used in an all-against-all analysis and a bit-wise matrix is created, '1' for significant similarity and '0' for no similarity between two proteins. To correct opposing results, e.g. if protein A matches protein B, but the reverse is false, a Smith and Waterman alignment is used (Smith and Waterman, 1981). Multiple domain proteins are detected by observing a case where protein B matches both protein A and C, but A and C do not match. To confirm that there is no significant similarity between A and C a Smith and Waterman alignment is again used. A recursive single linkage clustering operation is then performed on the constructed matrix to obtain a similarity table and a table of cluster assignments. Multidomain proteins can be shared between multiple clusters without introducing negative relationships, therefore the exact domain boundaries are not required.

Delineating Domains Using Physical Principles

There have been many algorithms developed for the assignment of domains by sequence comparison, but there are only a few available methods for the assignment of structural domains based on physical principles alone. Despite the creativity in these early approaches (Busetta and Barrans, 1984; Vonderviszt *et al.*, 1986; Kikuchi *et al.*, 1988), none appeared successful in providing reliable domain boundary predictions.

Recently, the method 'Domain Guess by Size (DGS)' has been described to predict domain boundaries based on the observation that domains have strict limits on their sequence size (Wheelan *et al.*, 2000). DGS is designed to calculate the likelihood of partitioning a sequence into one or more domains using a distribution of domain sizes derived from a set of proteins of known 3D structure. It is very accurate at finding the domain boundaries within two domain proteins, but it fails when proteins have more than two domains.

New Approaches to Domain Boundary Prediction

Analysis of Inter-domain Linkers and Their Prediction

Recent studies have shown that linkers connecting domains play an essential role in maintaining cooperative inter-domain interactions (Gokhale and Khosla, 2000). Altering the length and amino acid content of linkers connecting domains has shown to have effects on protein stability, folding rates and domain-domain orientation (van Leeuwen *et al.*, 1997; Robinson and Sauer, 1998).

Inter-domain linkers can be split into two classes, helical and non-helical. The compositions of helical and non-helical linkers are different from each other and are also distinct from the composition of helices, strands and loops within domains. Inter-domain linkers are distinct from loops within domains: a recent study (Crasto and Feng, 2001) showed residues Pro, Gly, Asp, Asn, His, Ser and Thr to be preferred in loop regions (in that order of abundance). We found that the residue types Gly, Asp, Asn, and Ser are avoided in linkers, and that His and Thr show no preference. The residue type Pro shows the highest preference in both loops and linkers, but however within a loop it is likely to be involved in a tight proline turn, whereas only a few such proline turns were observed in linkers.

Pro is the most prefered residue type within non-helical linkers. It is likely to be favoured because it has no amide hydrogen to donate in hydrogen bonding and, therefore, structurally isolates the linker from the domains. Pro often forms a cis-Pro isomer, which reverses the direction of the polypeptide chain. The latter could well lead to clashes of two domains during folding and formation of non-native interactions. Therefore, residues found to precede Pro in linkers often show relatively fast cis-trans conversion kinetics and at the same time exhibit a preference for the trans conformation. A helical linker might also be important for correct domain folding, as it could act as a rigid spacer to separate two domains. Helical conformations are rapidly formed during folding, well before the native structure is established (Aurora *et al.*, 1997), allowing the domains to fold independently without forming non-native interactions.

These observations suggest that a successful method to predict linkers may come from the amino acid propensities held within the two linker types. We tried various sliding window and neural network approaches to exploit amino acid composition and dipeptide patterns. Unfortunately, our techniques do not yet surpass 25% accuracy in direct linker prediction, which is too low to be used in any domain prediction scheme. Nonetheless, we have compiled a linker database which is available at http: //mathbio.nimr.mrc.ac.uk. The database is organized into six categories, based on the number of linkers between two domains, 1-linkers, 2-linkers, 3-linkers, greater than 3-linkers, and the two classes 'non-helical' and 'helical'. The database can be searched using several query types, such as PDB code, PDB header, linker length, C^{α} extent or sequence. Searches using regular expressions are possible and can be used to search for particular sequence motifs.

Predicting Domain Boundaries Based on Hydrophobic Collapse

Although many models of protein folding have been described (Dill *et al.*, 1995), the main driving force of protein folding was recognised early on to

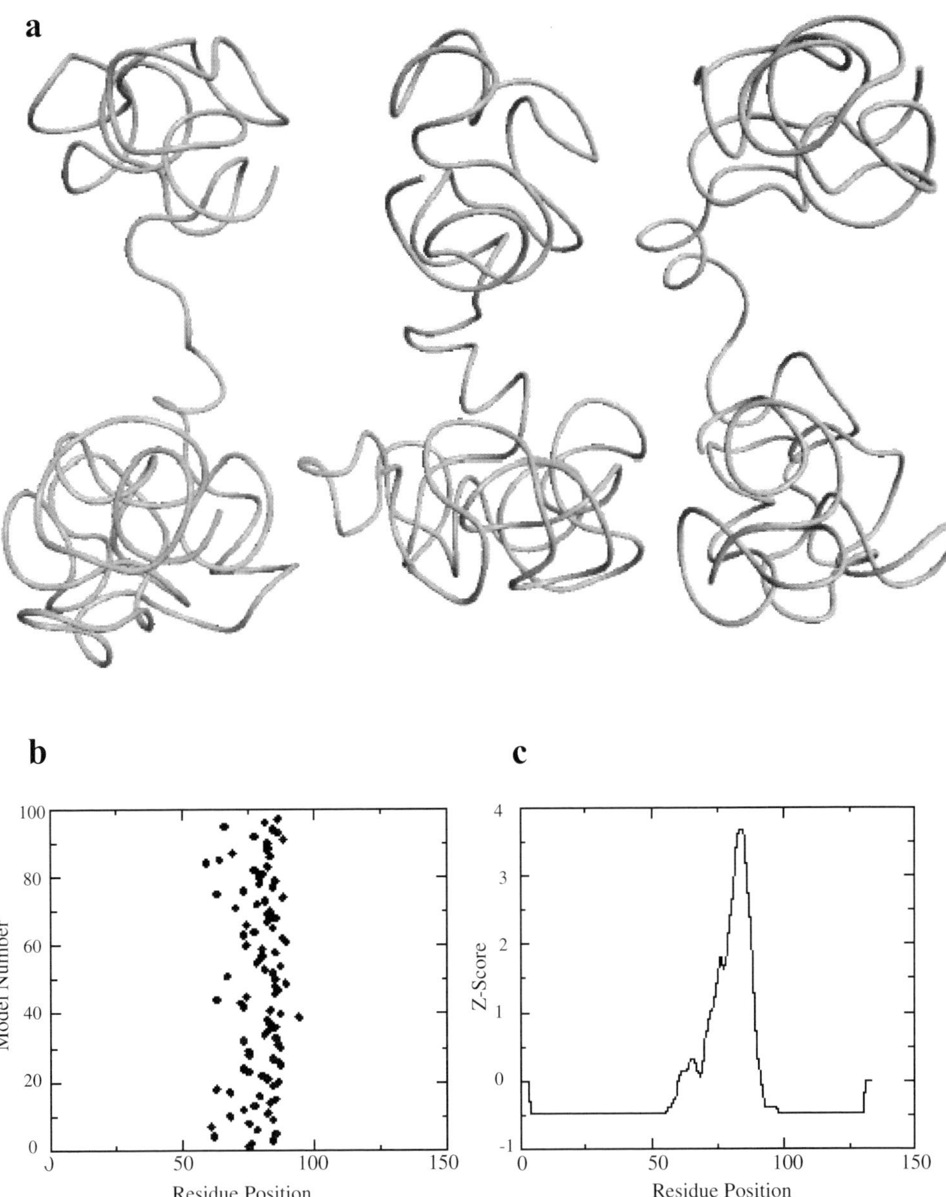

Figure 2. Overview of the SnapDRAGON method as applied to protein 2eifA: a) Three example DRAGON models visualised using the MOLMOL package (Koradi *et al.*, 1996). b) Domain boundary positions for 100 DRAGON models. c) Smoothing of boundary positions, the peak represents a predicted boundary. Figure adapted with permission from George and Heringa, 2002a.

be the burial of hydrophobic side chains into the interior of the molecule, creating a hydrophobic core and hydrophilic surface (Kauzmann, 1959). Since then, many authors have stressed the importance of the hydrophobic collapse for protein folding (Rackovsky and Scheraga, 1977; Dill, 1985). Since hydrophobic interactions play a major role in organising and stabilising the tertiary structure of proteins (Perutz *et al.*, 1965; Rose *et al.*, 1985) such interactions are seen to be preserved in fold families where there is little or no sequence homology between its members (Lesk and Chothia, 1980). Following this notion we have developed a protein folding method that is based on the clustering of hydrophobic residues in space in an attempt to predict domain boundaries. Our method SnapDRAGON (George and Heringa, 2002a) folds a polypeptide primarily based on the notion of conserved hydrophobicity of amino acids, as well as secondary structure prediction. In principle, SnapDRAGON employs the DRAGON algorithm (Aszódi and Taylor, 1994; Aszódi *et al.*, 1995; Aszódi and Taylor, 1997) to generate a large number of *ab initio* 3D model structures for a given multiple sequence alignment, with predicted secondary structure, and automatically assigns the domain boundaries for each of the models. A final prediction of the domain boundaries is derived from the consistency of the domain boundary assignments observed in the set of alternative 3D models.

The models vary considerably in structure with different domain contents and associated boundary positions. However, at this stage we are not interested in the details of the overall fold, but merely if we can consistently form isolated globular units given a multiple alignment and an idea of its secondary structure.

In Figure 2 an outline of the SnapDRAGON method is given. The figure shows three DRAGON models out of a total of 100 models normally generated (Figure 2a). Figure 2b shows the boundary positions assigned automatically to each of the 100 DRAGON models by the method of Taylor (1999). The boundaries for each DRAGON model are then summed along the length of the sequence and smoothed using a biased window approach. The resulting boundary scores are finally normalised according to a protocol to derive statistical significance. Each significant peak in the graph represents a predicted boundary (Figure 2c).

The overall accuracy of domain content prediction for a dataset of 183 singular domain proteins and 231 multiple domain proteins is 72.4% (±17.4%). All domain boundary predictions were compared to the linker positions in the real 3D protein structures. Overall, linker prediction accuracy by SnapDRAGON is 51.8% (±39.1%) per protein over the 231 multidomain proteins.

DOMAINATION: Domain Identification by Comparative Sequence Analysis
To enhance the identification of domain boundaries using sequence database homology searching, we have developed a method called DOMAINATION (George and Heringa, 2002b). The method currently utilises PSI-BLAST version 2.0.10 to identify homologues to a query sequence. The distribution of the aligned positions of N- and C-terminus from PSI-BLAST local sequence alignments is used to identify potential domain boundaries. DOMAINATION incorporates a new strategy for chopping and joining domains and domain segments in an attempt to track a protein's evolutionary pathway from its loss and gain of domains. Profiles are created for each domain or domains inferred from the corresponding PSI-BLAST local alignments and used in further database iterative searches.

Database Search Protocol
The method begins with an initial run of PSI-BLAST (Altschul *et al.*, 1997) to search the non-redundant sequence databases (NRDB) with a single query sequence. All significant gapped local alignments are collected from the PSI-BLAST output. A new domain cutting protocol is then applied to the query sequence, based on the distribution of N- and C- termini of the local alignments. Multiple sequence alignments are generated in parallel for each domain and a non-redundant set of corresponding local alignments. Selection of sequences for the alignment is achieved using OBSTRUCT (Heringa *et al.*, 1992) to find the largest subset of sequences within a range of 20% to 60% sequence identity. Each domain sequence set must contain the original query sequence to prevent profile wander, which occurs when the search strategy is too permissive such that false-positives are attracted in the profile, resulting in the possible loss of truly homologous sequences found in earlier rounds. Multiple alignments are constructed for each domain sequence set using PRALINE (Heringa, 1999). The multiple alignments are submitted simultaneously in further PSI-BLAST searches of the NRDB. The whole process iterates until no new sequences are detected or when domain cutting ends.

Assigning Continuous and Discontinuous Domains
Across an alignment resulting from a PSI-BLAST run, the numbers of N- and C-termini are counted along the length of the query. The two N- and C-termini distributions are then combined using a window approach. Higher weight is assigned to regions that have an abundance of both N- and C-termini, as these are likely to represent a domain boundary. A new element in DOMAINATION is that it attempts to detect discontinuous domains by

calculating an 'independence' score for each segment. Segment independence is based on the percentage of sequences that align with the segment and not with any other, where it is insisted that a matching sequence overlaps the segment by at least 70%. A segment is considered independent if more than 10% of its matching sequences do not align with any other segment. Such a region is termed a homology domain, and it is thought to be able to exist independently.

Any segment with an independence score below 10% is joined to other 'dependent' segments in the following way: if a number of sequences match two non-adjacent segments (>70% overlap to each) and do not match the segment(s) in between (<30%), then these segments are likely to form parts of a discontinuous domain. Segments are joined if the majority (>50%) of sequences have association between the two segments. With this approach segments constituting discontinuous domains can be identified. Many proteins predicted to have three or more domains often have joining events in the first iteration of DOMAINATION. This occurs when a central domain is observed to have been inserted between two 'parent' domains at some stage of the protein's evolution. DOMAINATION detects the insertion and joins the parent domains together. After subsequent database searches the parent domains are then separated and submitted as individual database searches. The ability to join fragments that flank an identified domain sets DOMAINATION aside from other methods that use PSI-BLAST to delineate domains (see above).

Analysis Of Domains Using Genomic Data

At present, there are nearly 100 published completed genome sequences. This plethora of sequence data requires automated annotation and comparisons between organisms. Proteins function in co-ordinated networks (Eisenberg *et al.*, 2000), but only a fraction of these networks have been identified and characterised through classical biochemistry, structural analysis, or activity assays. In order to extract the maximum amount of information from genomic data, all conserved genes need to be classified according to their sequential relationships (Tatusov *et al.*, 1997; Russell and Ponting, 1998).

Phylogenetic profiling is a means for studying genes that have the same pattern of presence or absence across multiple species, revealing sets of proteins that have coevolved, and are therefore likely to act in the same cellular process (Pellegrini *et al.*, 1999). The 'Rosetta Stone method' is based on the observation that individual domains of a multidomain protein in one

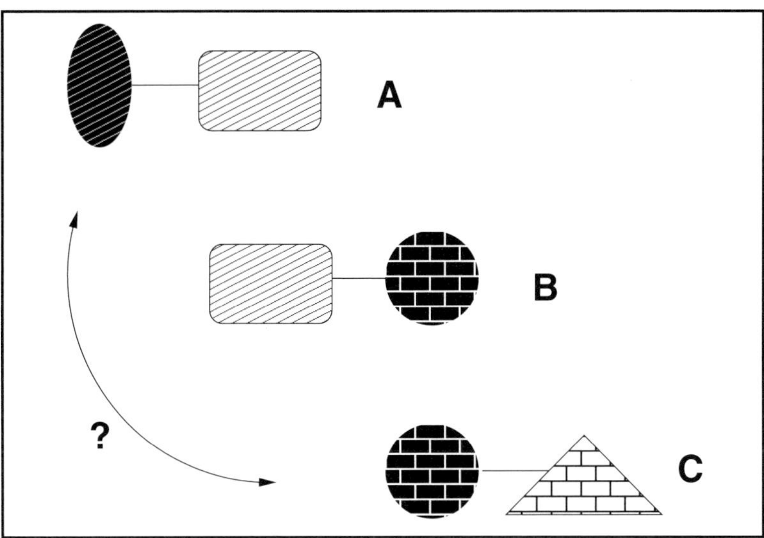

Figure 3. Illustration of false sequence clustering. Three dual-domain proteins A, B, and C with homologies between single domains in pairs A/B and B/C would be clustered together by sequence search without taking precaution, although no direct relationship exists between A and C.

organism are present as separate single-domain proteins in a second organism (Marcotte *et al.*, 1999). The existence of the multidomain protein implies that the separate proteins interact in the second organism.

The organisation of sequences into families is important to provide evolutionary, functional and structural data, and to organise sequential data into a manageable database. In cases where the proteins consist of several domains, errors within families can occur. For example, three dual-domain proteins A, B and C are cartooned in Figure 3. Owing to the homology of single domains in the pairs A/B and B/C, all three proteins would be clustered into the same family, although proteins A and C are not related. A and C are incorrectly associated through homology to protein B. Therefore, domain identification is essential for accurate clustering of sequences.

Conclusion

Over the last few years many protein domain databases and tools to recognise and delineate domains have been constructed. However, there is still considerable inconsistency between the annotations so that efforts to streamline the data, such as initiated by the InterPro consortium, should lead to increased quality of the data. This in turn should improve benchmarking required for each individual development and foster a thorough comparison of prediction methods. Another promising avenue would be the integration of various prediction methods in a common interface, leading to improved prediction. Despite the recent improvements in domain delineation outlined in this chapter, it is clear that further developments will be sorely needed for the accurate annotation of genome sequences.

References

Adams, R., Das, S., and Smith, T. 1996. Multiple domain protein diagnostic patterns. Protein Sci. 5: 1240-1249.

Altschul, S., Gish, W., Miller, W., Myers, E., and Lipman, D. 1990. Basic local alignment search tool. J. Mol. Biol. 214: 403-410.

Altschul, S., Madden, T., Schaffer, A., Zhang, J., Zhang, Z., Miller, W., and Lipman, D. 1997. Gapped BLAST and PSI-BLAST: a new generation of protein database search programs. Nucleic Acids Res. 25: 3389-402.

Apic, G., Gough, J., and Teichmann, S. 2001. Domain combinations in archaeal, eubacterial and eukaryotic proteomes. J. Mol. Biol. 310: 311-325.

Apweiler, R., Attwood, T., Bairoch, A., Bateman, A., Birney, E., Biswas, M., Bucher, P., Cerutti, L., Corpet, F., Croning, M., Durbin, R., Falquet, L., Fleischmann, W., Gouzy, J., Hermjakob, H., Hulo, N., Jonassen, I., Kahn, D., Kanapin, A., Karavidopoulou, Y., Lopez, R., Marx, B., Mulder, N., Oinn, T., Pagni, M., Servant, F., Sigrist, C., and Zdobnov, E. 2000. InterPro - an integrated documentation resource for protein families, domains and functional sites. Bioinformatics 16: 1145-1150.

Aszódi, A., Gradwell, M., and Taylor, W. 1995. Global fold determination from a small number of distance restraints. J. Mol. Biol. 251: 308-326.

Aszódi, A. and Taylor, W. 1994. Folding polypeptide α-carbon backbones by distance geometry methods. Biopolymers 34: 489-506.

Aszódi, A. and Taylor, W. 1997. Hierarchical Inertial Projection: a fast distance matrix embedding algorithm. Computers Chem. 21: 13-23.

Attwood, T., Flower, D., Lewis, A., Mabey, J., Morgan, S., Scordis, P., Selley, J., and Wright, W. 1999. PRINTS prepares for the new millennium. Nucleic Acids Res., 27: 220-225.

Aurora, R., Creamer, T., Srinivasan, R., and Rose, G. 1997. Local interactions in protein folding: lessons from the alpha-helix. J. Biol. Chem. 272: 1413-1416.

Bairoch, A. and Apweiler, R. 2000. The SWISS-PROT protein sequence database and its supplement TrEMBL in 2000. Nucleic Acids Res. 28: 45-48.

Bateman, A., Birney, E., Durbin, R., Eddy, S., Howe, K., and Sonnhammer, E. 2000. The Pfam protein families database. Nucleic Acids Res. 28: 263-266.

Bennett, M., Schlunegger, M., and Eisenberg, D. 1995. 3D domain swapping: a mechanism for oligomer assembly. Protein Sci. 4: 2455-2468.

Bonneau, R., Strauss, C., and Baker, D. 2001. Improving the performance of rosetta using multiple sequence alignment information and global measures of hydrophobic core formation. Proteins 43: 1-11.

Bork, P. 1991. Shuffled domains in extracellular proteins. FEBS Lett. 286: 47-54.

Bork, P., Sander, C., and Valencia, A. 1992. An ATPase domain common to prokaryotic cell cycle proteins, sugar kinases, actin, and hsp70 heat shock proteins. Proc. Nat. Acad. Sci. USA 89: 7290-7294.

Busetta, B. and Barrans, Y. 1984. The prediction of protein domains. Biochim. Biophys. Acta 790: 117-124.

Campbell, I. and Downing, A. 1994. Building protein structure and function from modular units. Trends Biotechnol. 12: 168-172.

Castiglone-Morelli, M., Stier, G., Gibson, T., Joseph, C., Musco, G., Pastore, A., and Trave, G. 1995. The KH module has an alpha beta fold. FEBS Lett. 358: 193-198.

Corpet, F., Servant, F., Gouzy, J., and Kahn, D. 2000. ProDom and ProDom-CG: tools for protein domain analysis and whole genome comparisons. Nucleic Acids Res. 28: 267-269.

Crasto, C. and Feng, J. 2001. Sequence codes for extended conformation: a neighbor-dependent sequence analysis of loops in proteins. Proteins 42: 399-413.

Crippen, G. 1978. The tree structural organisation of proteins. J. Mol. Biol. 126: 315-332.

Das, S. and Smith, T. 2000. Identifying nature's protein LEGO set. Adv. Protein Chem. 54: 159-183.

Davidson, J., Chen, K., Jamison, R., Musmanno, L., and Kern, C. 1993. The evolutionary history of the first three enzymes in pyrimidine biosynthesis. Bioessays 15: 157-164.

Dietmann, S., Park, J., Notredame, C., Heger, A., Lappe, M., and Holm, L. 2001. A fully automatic evolutionary classification of protein folds: Dali Domain Dictionary version 3. Nucleic Acids Res. 29: 55-57.

Dill, K. 1985. Theory for the folding and stability of globular proteins. Biochemistry 24: 1501-1509.

Dill, K., Bromberg, S., Yue, K., Fiebig, K., Yee, D., Thomas, P., and Chan, H. S. 1995. Principles of protein folding - a perspective from simple exact models. Protein Sci. 4: 561-602.

Doolittle, R. 1995. The multiplicity of domains in proteins. Annu. Rev. Biochem. 64: 287-314.

Eisenberg, D., Marcotte, E., Xenarios, I., and Yeates, T. 2000. Protein function in the post-genomic era. Nature 405: 823-826.

Enright, A. and Ouzounis, C. 2000. GeneRAGE: a robust algorithm for sequence clustering and domain detection. Bioinformatics 16: 451-457.

Garel, J. 1992. Folding of large proteins: Multidomain and multisubunit proteins. In: Protein Folding. 1st edition. Creighton,T., ed. W.H. Freeman and Company, New York. p. 405-454.

George, D., Hunt, L., and Barker, W. 1996. PIR-international protein sequence database. Method. Enzymol. 266: 41-59.

George, R., and Heringa, J. 2002a SnapDRAGON: a method to delineate protein structural domains from sequence data. J. Mol. Biol. 316: 839-851.

George, R., and Heringa, J. 2002b. Protein domain identification and improved sequence similarity searching using PSI-BLAST. Protein Struct. Funct. Genet. 48: 672-681.

Go, M. 1978. Correlation of DNA exonic regions with protein structural units in haemoglobin. Nature 291: 90-92.

Gokhale, R. and Khosla, C. 2000. Role of linkers in communication between protein modules. Curr. Opin. Chem. Biol. 4: 22-27.

Gouzy, J., Corpet, F., and Kahn, D. 1999. Whole genome protein domain analysis using a new method for domain clustering. Comput. Chem. 23: 333-340.

Gracy, J. and Argos, P. 1998a. Automated protein sequence database classification. II. Delineation of domain boundaries from sequence similarities. Bioinformatics 14: 174-87.

Gracy, J. and Argos, P. 1998b. DOMO: a new database of aligned protein domains. Trends Biochem. Sci. 23: 495-497.

Guerois, R. and Serrano, L. 2001. Protein design based on folding models. Curr. Opin. Struct. Biol. 11: 101-106.

Hadley, C. and Jones, D. 1999. A systematic comparison of protein structure classifications SCOP, CATH and FSSP. Structure Fold. Des. 7: 1099-1112.

Henikoff, J. G., Greene, E., Pietrokovski, S., and Henikoff, S. 2000. Increased coverage of protein families with the blocks database servers. Nucleic Acids Res. 28: 228-230.

Henikoff, S., Greene, E., Pietrokovski, S., Bork, P., Attwood, T., and Hood, L. 1997. Gene families: the taxonomy of protein paralogs and chimeras. Science 278: 609-614.

Heringa, J. 1999. Two strategies for sequence comparison: profile-preprocessed and Secondary structure-induced multiple alignment. Comput. Chem. 23: 341-364.

Heringa, J., Sommerfeldt, H., Higgins, D., and Argos, P. 1992. OBSTRUCT: a program to obtain largest cliques from a protein sequence set according to structural resolution and sequence similarity. Comput Appl. Biosci. 8: 599-600.

Heringa, J. and Taylor, W. 1997. Three-dimensional domain duplication, swapping and stealing. Curr. Opin. Struct. Biol. 7: 416-421.

Hofmann, K., Bucher, P., Falquet, L., and Bairoch, A. 1999. The PROSITE database, its status in 1999. Nucleic Acids Res. 27: 215-219.

Holm, L. and Sander, C. 1994. Parser for protein folding units. Proteins 19: 256-268.

Holm, L. and Sander, C. 1997. Dali/FSSP classification of three-dimensional protein folds. Nucleic Acids Res. 25: 231-234.

Islam, S., Luo, J., and Sternberg, M. 1995. Identification and analysis of domains in proteins. Prot. Eng. 8: 513-525.

Jacob, F. 1977. Evolution and tinkering. Science 196: 1161-1166.

Janin, J. and Wodak, S. 1983. Structural domains in proteins an their role in the dynamics of protein function. Prog. Biophys. Molec. Biol. 42: 21-78.

Jones, S., Stewart, M., Michie, A., Swindells, M., Orengo, C., and Thornton, J. 1998. Domain assignment for protein structures using a consensus approach: characterization and analysis. Prot. Sci. 7: 233-242.

Kauzmann, W. 1959. Some factors in the interpretation of protein denaturation. Adv. Prot. Chem. 14: 1-63.

Kikuchi, T., Nemethy, G., and Scheraga, H. 1988. Prediction of the location of structural domains in globular proteins. J. Prot. Chem. 7: 427-71.

Koradi, R., Billeter, M., and Wüthrich, K. 1996. MOLMOL: a program for display and analysis of macromolecular structures. J. Mol. Graph. 14: 51-55.

Kuroda, Y., Tani, K., Matsuo, Y., and Yokoyama, S. 2000. Automated search of natively folded protein fragments for high-throughput structure determination in structural genomics. Protein Sci. 9: 2313-2321.

Lesk, A. and Chothia, C. 1980. How different amino acid sequences determine similar protein structures: The structure and evolutionary dynamics of the globins. J. Mol. Biol. 136: 225-270.

Marcotte, E., Pellegrini, M., Ng, H., Rice, D., Yeates, T., and Eisenberg, D. 1999. Detecting protein function and protein-protein interactions from genome sequences. Science 285: 751-753.

Murvai, J., Vlahovicek, K., Barta, E., Cataletto, B., and Pongor, S. 2000. The SBASE protein domain library, release 7.0: a collection of annotated protein sequence segments. Nucleic Acids Res. 28: 260-262.

Murzin, A., Brenner, S., Hubbard, T., and Chothia, C. 1995. SCOP: a structural classification of proteins database for the investigation of sequences and structures. J. Mol. Biol. 247: 536-540.

Nielsen, P. and Yamada, Y. 2001. Identification of cell-binding sites on the Laminin α5 N-terminal domain by site-directed mutagenesis. J. Biol. Chem. 276: 10906-109012.

Orengo, C., Jones, D., and Thornton, J. 1994. Protein superfamiles and domain superfolds. Nature 372: 631-634.

Orengo, C., Michie, A., Jones, S., Jones, D., Swindells, M., and Thornton, J. 1997. CATH - a hierarchic classification of protein domain structures. Structure 5: 1093-1108.

Park, J. and Teichmann, S. 1998. DIVCLUS: an automatic method in the GEANFAMMER package that finds homologous domains in single- and multi-domain proteins. Bioinformatics 14: 144-150.

Pellegrini, M., Marcotte, E., Thompson, M., Eisenberg, D., and Yeates, T. 1999. Assigning protein functions by comparative genome analysis: protein phylogenetic profiles. Proc. Natl. Acad. Sci. USA 96: 4285-4288.

Perutz, M., Kendrew, J., and Watson, H. 1965. Structure and function of heamoglobin. II. some relations between polypeptide chain configuration and amino acid sequence. J. Mol. Biol. 13: 669-678.

Pfuhl, M. and Pastore, A. 1995. Tertiary structure of an immunoglobulin-like domain from the giant muscle protein titin: a new member of the I set. Structure 3: 391-401.

Rackovsky, S. and Scheraga, H. 1977. Hydrophobicity, hydrophilicity, and the radial and orientational distributions of residues in native proteins. Proc. Natl. Acad. Sci. USA 74: 5248-5251.

Robinson, C. and Sauer, R. 1998. Optimizing the stability of single-chain proteins by linker length and composition mutagenesis. Proc. Nat. Acad. Sci. USA 95: 5929-5934.

Rose, G., Geselowitz, A., Lesser, G., Lee, R., and Zehfus, M. 1985. Hydrophobicity of amino acid residues in globular proteins. Science 229: 834-838.

Rose, G. D. 1979. Hierarchic organisation of domains in globular proteins. J. Mol. Biol. 234: 447-470.

Rossmann, M., Moras, D., and Olsen, K. 1974. Chemical and biological evolution of nucleotide binding proteins. Nature 250: 194-199.

Russell, R. and Ponting, C. 1998. Protein fold irregularities that hinder sequence analysis. Curr. Opin. Struct. Biol., 8: 364-371.

Schultz, J., Copley, R., Doerks, T., Ponting, C., and Bork, P. 2000. SMART: a web-based tool for the study of genetically mobile domains. Nucleic Acids Res. 28: 231-234.

Siddiqui, A. and Barton, G. 1995. Continuous and discontinuous domains - an algorithm for the automatic generation of reliable protein domain definitions. Prot. Sci. 4: 872-884.

Siddiqui, A., Dengler, U., and Barton, G. 2001. 3Dee: a database of protein structural domains. Bioinformatics 17: 200-201.

Smith, R. and Smith, T. 1992. Pattern-induced multi-sequence alignment (PIMA) algorithm employing secondary structure-dependent gap penalties for use in comparative protein modelling. Protein Eng. 5: 35-41.

Smith, T. and Waterman, M. 1981. Identification of common molecular subsequences. J. Mol. Biol. 147: 195-197.

Sonnhammer, E. and Durbin, R. 1994. A workbench for large-scale sequence homology analysis. Comput. Appl. Biosci. 10: 301-307.

Srinivasarao, G., Yeh, L., Marzec, C., Orcutt, B., Barker, W. C., and Pfeiffer, F. 1999. Database of protein sequence alignments: PIR-ALN. Nucleic Acids Res. 27: 284-285.

Swindells, M. 1995. A procedure for detecting structural domains in proteins. Prot. Science 4: 103-112.

Tatusov, R., Koonin, E., and Lipman, D. 1997. A genomic perspective on protein families. Science 278: 631-637.

Tatusov, R., Natale, D., Garkavtsev, I., Tatusova, T., Shankavaram, U., Rao, B. S., Kiryutin, B., Galperin, M., Fedorova, N., and Koonin, E. 2001. The COG database: new developments in phylogenetic classification of proteins from complete genomes. Nucleic Acids Res. 29: 22-28.

Taylor, W. 1999. Protein structure domain identification. Prot. Eng. 12: 203-216.

Taylor, W. and Orengo, C. 1989. A holistic approach to protein structure comparison. Prot. Eng. 2: 505-519.

Tsai, C. J. and Nussinov, R. 1997. Hydrophobic folding units derived from dissimilar monomer structures and their interactions. Protein Sci. 6: 24-42.

van Leeuwen, H., Strating, M., Rensen, M., deLaat, W., and van der Vliet, P. 1997. Linker length and composition influence the flexibility of Oct-1 DNA binding. EMBO J. 16: 2043-2053.

Vonderviszt, F., Matrai, G., and Simon, I. 1986. Characteristic sequential residue environment of amino acids in proteins. Int. J. Pept. Protein Res. 27: 483-492.

Wheelan, S., Marchler-Bauer, A., and Bryant, S. 2000. Domain size distributions can predict domain boundaries. Bioinformatics 16: 613-618.

Zehfus, M. 1997. Identification of compact, hydrophobically stabilized domains and modules containing multiple peptide chains. Protein Sci. 6: 1210-1219.

From: *Bioinformatics and Genomes: Current Perspectives*
Edited by: Miguel A. Andrade

Chapter 2

Using Multiple Alignment Methods to Assess the Quality of Genomic Data Analysis

Cédric Notredame and Chantal Abergel

ABSTRACT

The analysis of multiple sequence alignments can generate essential clues for genomic data analysis. Yet, to be informative such analyses require some means of estimating the reliability of a multiple alignment. In this chapter we describe a novel method allowing the unambiguous identification of the residues correctly aligned within a multiple alignment. This method uses an index named CORE (Consistency of the Overall Residue Evaluation) based on the T-Coffee multiple sequence alignment algorithm. We provide two examples of applications: one where the CORE index is used to identify correct blocks within a difficult multiple alignment and another where the CORE index is used on genomic data to identify the proper start codon and a frame-shift within one of the sequences.

INTRODUCTION

Biological analysis largely relies on the assembly of elaborate models meant to summarize our knowledge of living complex mechanisms. For that purpose, vast amounts of data are collected, analyzed, validated and then integrated within a model. In an ideal world, an existing model would be available to explain every bit of experimental data. In the real world, this is rarely the case, and every day, existing models need to be modified to accommodate new findings. Sometimes, data that cannot be explained is kept at bay until the accumulation of new evidences prompts the design of an entirely new model. Unaccountable data can be viewed as the stuff inflating an inconsistency bubble. Eventually, the bubble bursts and a new model is designed.

A multiple alignment is nothing less than such a model. Given a series of sequences and an alignment criteria (structure similarity, common phylogenetic origin) the multiple alignment contains a series of hypotheses regarding the relationship between the sequences it is made of. This alignment can accommodate data generated experimentally (e.g. alignment of two homologous catalytic residues) or combine the results of various sequence analysis methods. The importance of the use of multiple sequence alignments in the context of sequence analysis has been recognized for a long time and it is so well established that most bioinformatics protocols make use of it. Multiple alignments have been turned into profiles (Gribskov *et al.*, 1987) and hidden Markov models (Krogh *et al.*, 1994) to enhance the sensitivity and the specificity of database searches (Altschul *et al.*, 1997). State of the art methods for protein structure prediction depend on the proper assembly of a multiple sequence alignment (Jones, 1999) as do phylogenetic analysis (Duret *et al.*, 1994). Over the last years multiple sequence alignment techniques have been instrumental to improvements made in almost every key area of sequence analysis. Yet, despite its importance, the accurate assembly of a multiple sequence alignment is a complex process, the biological knowledge and the computational abilities it requires are far beyond our current capacities. As a consequence, biologists are left to use approximate programs that attempt to assemble proper alignments without providing any guaranty they may do so. The lack of a 'perfect' or at least reasonably robust method explains why so many multiple sequence alignment packages exist. The variations among these packages are not only cosmetic; they include the use of very different algorithms, different parameters and generally speaking different paradigms. For a recent review of state-of-the-art techniques, see (Duret and Abdeddaim, 2000).

Database searches, structure predictions, phylogenetic analysis are enough on their own to make multiple alignment compulsory in a genome analysis task. Yet, thanks to the sanity checks they provide, multiple alignments can also be instrumental at tackling the plague of genomic analysis: faulty data. When dealing with genomes, faulty data arises from two major sources: sequencing errors and wrong predictions. The consequence is that a predicted protein sequence may have accumulated errors both at the DNA level and when its frame was predicted (this will be especially true in eukaryotic genes where exons may be missed, added or improperly predicted). In the worst cases, the effect of such errors will be amplified in the high level analysis, leading to an improper analysis of the available data. On the other hand, once they have been identified, these errors are usually easily corrected either by extra sequencing or data extrapolation. Therefore, any method providing a reasonable sanity-check that earmarks areas of a genome likely to be problematic would be a major improvement. In this chapter we will show how multiple sequence alignments can be used to carry out part of this task. For that purpose we will focus on the applications of T-Coffee, a recently described method (Notredame *et al.*, 2000).

Generating Multiple Alignments With T-Coffee

Despite the large variety of multiple sequence alignment methods publicly available, the number of packages effectively used for data analysis is surprisingly small and a vast majority of the alignments found in the literature are produced using only two programs: ClustalW (Thompson *et al.*, 1994) and its X-Window implementation ClustalX. ClustalW uses the progressive alignment strategy described by Taylor (Taylor, 1988) and Doolitle (Feng and Doolittle, 1987), refined in order to incorporate sequence weights and a local gap penalty scheme. Recently, the ClustalW algorithm was further modified in order to improve the accuracy of the produced alignments by making the evaluation of the substitution costs position dependent. This improved algorithm is implemented in the T-Coffee package (Notredame *et al.*, 2000).

The aim of T-Coffee is to build a multiple alignment that has a high level of consistency with a library of pre-computed pair-wise alignments. This library may contain as many alignments as one wishes and it may also be redundant and inconsistent with itself. For instance it may contain several alternative alignments of the same sequences aligned using various gap penalties. It may also contain alternative alignments obtained by applying different methods onto the sequences. Overall, the library is a collection of alignments believed to be correct. Within this library, each alignment receives a weight that is an estimation of its biological likeliness (i.e., how much trust does one have in this alignment to be correct). For that purpose, one may use any suitable criteria such as percent identity, P-Value estimation or any other appropriate method. The T-Coffee algorithm uses this library in order to compute the score for aligning two residues with one another in the multiple alignment. This score is named the extended weight because it requires an extension of the library. The extended weight takes into account the compatibility of the alignment of two residues with the rest of the alignments observed within the library, its derivation is extensively described in (Notredame *et al.*, 2000). The principle is straightforward: in order to compute the extended weight associated with two residues R and S of two different sequences, one will consider whether when R is found aligned in the library with some residue X of a third sequence, S is also found aligned with that same residue X in another entry of the library. If that is the case, then the weight associated with R and S will be increased by the minimum of the two weights RX and SX. The final extended weight will be obtained when every possible X has been considered and the resulting contributions summed up. Although this operation seems to be very expensive from a computational point of view, its effective computational cost is kept low thanks to the scarceness of the primary library (i.e., for most pairs of residues RS, very few Xs need to be considered). In the end, a pair of residues is highly consistent (and has a high extended weight) if most of the other sequences contain at least one residue that is found aligned both to R and to S in two different pair-wise alignments. A key property of this weight extension procedure is to concentrate information: the extended score of RS incorporates some information coming from all the sequences in the set and not only from the two sequences contributing R and S.

The main advantage of the extended weights is that they can be used in place of a substitution matrix. While standard substitution matrices do not discriminate between two identical residues (e.g. all the cysteins are the same for a Pam (Dayhoff *et al.*, 1979) or a Blosum substitution matrix (Henikoff and Henikoff, 1992)), the extended weights are truly position specific and make it possible to discriminate between two identical residues that only differ by their positions. Once the library has been assembled (potential ways of assembling that library are described later) and the extended weights computed, T-Coffee closely follows the ClustalW procedure using the extended weights instead of a substitution matrix. The overall T-Coffee strategy is outlined in Figure 1. All the sequences are first aligned two by two, using dynamic programming (Needleman and Wunsch, 1970) and the extended library in place of a substitution matrix. The distance matrix thus obtained is then used to compute a neighbor-joining tree (Saitou and Nei,

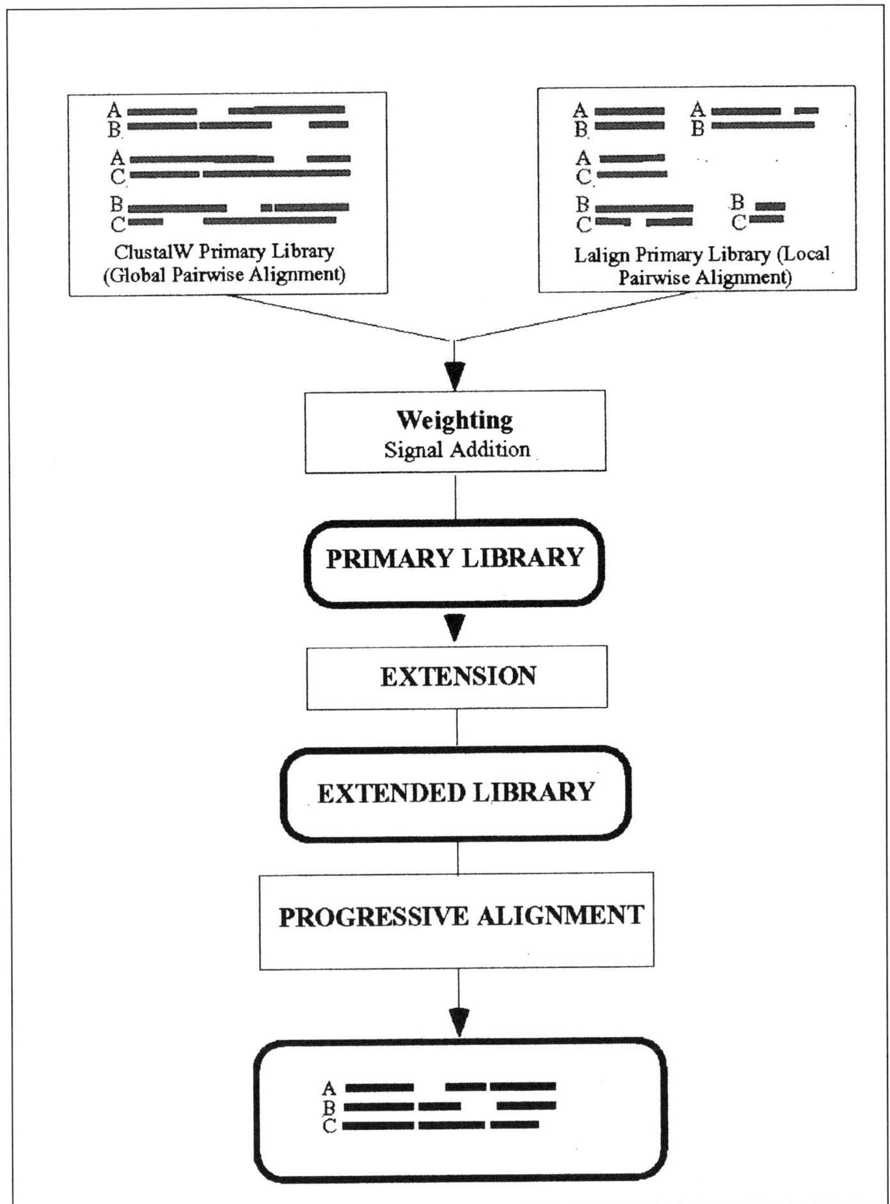

Figure 1. Layout of the T-Coffee algorithm. This figure indicates the chain of events that lead from unaligned sequences to a multiple sequence alignment using the T-Coffee algorithm. Data processing steps are boxed while data structures are indicated by rounded boxes.

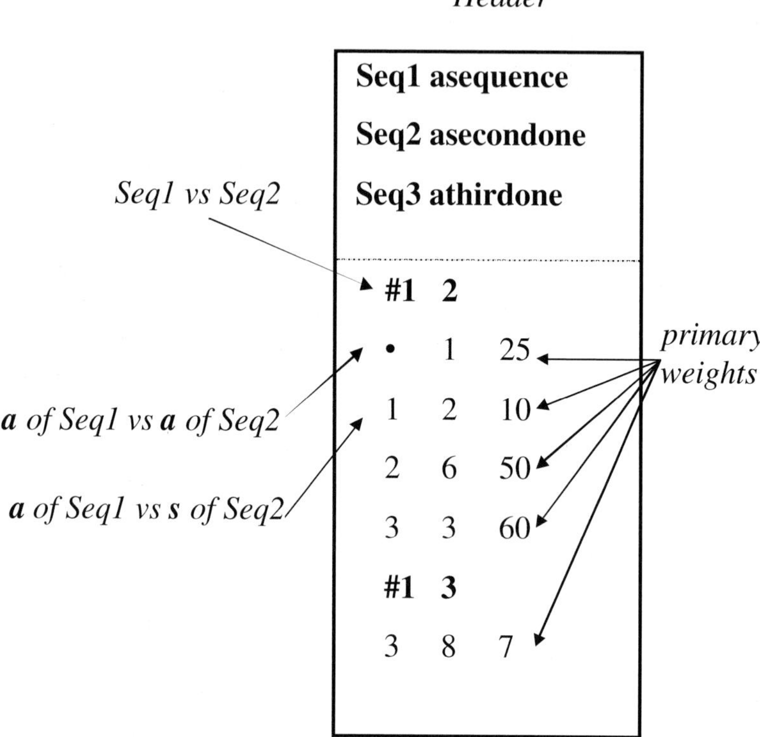

Figure 2. Library Format. An example of a library used by T-Coffee. The header contains the sequences and their names. '# 1 2' indicates that the following pairs of residues correspond to sequences 1 and 2. Each pair of aligned residues contains three values: the index of residue 1, the index of residue 2 and the weight associated with the alignment of these two residues. No order or consistency is expected within the library.

1987). This tree guides the progressive assembly of a multiple sequence alignment: the two closest sequences are first aligned by normal dynamic programming using the extended weights to align the residues in the two sequences; no gap penalty is applied (because it has already been applied to generate the alignments contained in the library). This pair of sequences is then fixed and any gaps that have been introduced cannot be shifted later. Then the program aligns the next closest two sequences or adds a sequence to the existing alignment of the first two sequences, depending which is suggested by the guide tree. The procedure always joins the next two closest sequences or pre-aligned group of sequences. This continues until all the sequences have been aligned. The extended weights are used, as before, to align two groups of pre-aligned sequences but taking the average library scores in each column of the existing alignments.

The key feature of T-Coffee is the freedom given to the user to build his own library following whatever protocol may seem appropriate. For this purpose, one may mix structural information with database results, knowledge-based information or pre-established collections of multiple alignments. It may also be necessary to explore a wide range of parameters given some computer package. A simple library format was designed to fit that purpose, as shown in Figure 2. A library is a straightforward ASCII file that contains a listing of every pair of aligned residue that needs to be described. Any knowledge-based information can easily be added manually to an automatically generated library or the other way round. The library can contain ambiguities and inconsistencies (i.e., two alignments possible for the first residue of Seq1 with Seq2). These ambiguities will be resolved while the alignment is being assembled, on the basis of the score given by the extended weights. The library does not need to contain a weight associated with each possible pair of residues. On the contrary, an ideal library only contains pairs that will effectively occur in the correct multiple alignment (i.e., N^2L pairs rather than N^2L^2 pairs, for an alignment of length L containing N sequences). While this flexibility to design and assemble one's own library is a very desirable property, in practice it is also convenient to have a standard automatic protocol available. Such a protocol exists and is fully integrated within the T-Coffee package. It is ran with the default mode and does not require the user to be aware of T-Coffee underlying concepts (library, extension, progressive alignment). This default protocol extensively described and validated in (Notredame *et al.*, 2000) requires two distinct libraries to be compiled and combined within the primary library before the extension. The first one contains a ClustalW pair-wise alignment of each possible pair of sequence within the dataset. For that purpose, ClustalW (Thompson *et al.*, 1994) is run using default parameters. This library is global because it is generated by aligning the sequences over their whole length (global alignments) using a linear space version of the Needleman and Wunsch algorithm (Needleman and Wunsch, 1970). The second library is local: for each possible pair of sequences, it contains the ten best non-overlapping local alignments as reported by the Lalign program (Huang and Miller, 1991) run with default parameters. In the local and the global libraries, each pair of residues found aligned is associated with a weight equal to the average level of identity within the alignment it came from. When a specific pair is found more than once, the weights associated with each occurrence are added. The main strength of this protocol is to combine local and global information within a multiple alignment. The level of consistency within the library will depend on the nature of the sequences. For instance, if the sequences are very diverse, the requirement for long insertions/deletions will often cause

Table 1

	cat 1	cat 2	cat 3	cat 4	cat 5	avg 1	avg 2
cw	79.53	32.91	48.72	74.02	67.84	67.89	61.82
prrp	78.62	32.45	50.14	51.12	82.72	66.45	60.25
dialign2	70.99	25.21	35.12	74.66	80.38	61.54	57.99
T-Coffee	**80.67**	**36.15**	**53.20**	**83.41**	**91.69**	**72.18**	**69.55**

To produce this table each dataset contained in the BaliBase was aligned using one of the methods (cw: ClustalW 1.81 (Thompson *et al.*, 1994), Prrp (Gotoh, 1996), dialign2 (Morgenstern *et al.*, 1998) and T-Coffee 1.29 (Notredame *et al.*, 2000). In BaliBase, reference alignments are classified in 5 categories: category 1 contains closely related sequences, category 2 contains a group of closely related sequences and an outsider, category 3 contains two groups of sequences that are distantly related, category 4 contains families with long internal indels, category 5 contains sequences with long terminal indels. The resulting alignments were then compared to their reference counterpart in BaliBase, only using the regions annotated as trustable in BaliBase. Under this scheme we define the accuracy of an alignment to be the percentage of columns that are found totally conserved in the reference divided by the total number of columns within that reference. The comparison is restricted to the portions annotated as trustworthy in the reference alignment. *avg 1* is the average of the results obtained on each of the 142 test cases, *avg 2* is the average of the results obtained in each category. Bold numbers indicate the best performing method.

the global alignments to be incorrect and inconsistent, while the local alignments will be less sensitive to that type of problems. In such a situation, the measure of consistence will enhance the local alignment signal and let it drive the multiple alignment assembly. Conversely, if the global alignments are good enough they will help removing the noise associated with the collection of local alignments (local alignments do not have any positional constraints). Overall, the current default T-Coffee protocol contains three distinct elements that lead to the collection of extended weights: the global library, the local library and the library extension that turns the sum of the two libraries into an extended library. Earlier work demonstrated that each of these components plays a significant part in improving the overall accuracy of the program. Table 1 shows that the current version of T-Coffee (Version 1.29) outperforms other popular multiple sequence alignment methods, as judged by comparison on BaliBase (Thompson *et al.*, 1999), a database of hand made reference structural alignments that are based on structural comparison (See Table 1 legend for a description of BaliBase and the comparison protocol).

These results illustrate well the good performances of T-Coffee on the wide range of situations that occur in BaliBase. It is especially interesting to point out that T-Coffee is the only method equally well suited to situations that require a global alignment strategy (categories 1, 2 and 3) and situations that are better served with a local alignment strategy (categories 4 and 5 with long internal and terminal insertions/deletions). The other methods are either good for global alignments (like ClustalW) or for local alignments (like Dialign2 (Morgenstern *et al.*, 1998)). It should be noted that T-Coffee still uses ClustalW 1.69 to construct the primary global library, because this was the last 'naïve' version of ClustalW, not tuned up on BaliBase. The latest version (1.81) has been tuned on the BaliBase references (hence its improved performances over the results originally reported for ClustalW). Using this ClustalW 1.81 version when benchmarking T-Coffee would make the process circular. Nonetheless, as good as it may seem, T-Coffee still suffers from the same shortcoming as any other package available today: *it is not always the best method.* Even if on average it does better than any of its counterparts, one cannot guaranty that T-Coffee will always generate the best alignment. For instance, although Dialign2 is significantly less good, it outperforms T-Coffee on 17 test sets (11%). ClustalW is better than T-Coffee in 24% of the cases. We may conclude from this that in practice, there will always be situations where some alternative method beats T-Coffee. Furthermore, even in cases where the T-Coffee improvement over any alternative method is very significant, it may lead to an alignment much less than 100% correct. This may not be so helpful since for practical usage, it would be much more helpful to know where the correctly aligned portions lie. This is so true that a method 20% correct and a proper estimation of its reliability would be much more useful than a method more accurate 'on average'.

Several situations exist in which a biologist can make use of this reliability information. For instance, if the purpose of the alignment is to extrapolate some experimental data onto an otherwise un-characterized genomic sequence, one will need to be very careful not to deduce anything from an unreliable portion of the alignment. More generally, unreliable positions within a multiple sequence alignment should not be used for predictive purpose. For instance, when turning a multiple alignment into a profile in order to scan databases for remote homologues, it is essential to exclude regions whose alignment cannot be trusted and that may obscure some otherwise highly conserved position. Used in this fashion, reliability information allows a significant decrease of the noise induced by locally spurious alignments. The other important application of a reliability measure is the identification of regions within a multiple alignment that are properly aligned without being highly conserved. These regions are extremely important when the alignment is used in conjunction with a predictive method that bases its analysis on mutation patterns. For instance, structure and phylogeny prediction methods require the presence of non-conserved positions to yield informative results. Any scheme that allows discriminating between positions that are degenerated but correctly aligned and positions that are simply misaligned may induce a dramatic improvement in the accuracy of these prediction methods. Furthermore a reliability measure will help identifying faulty data and provide some clues on how to correct it. In the next section, we show how consistency can be measured on a T-Coffee alignment and how that measure provides a fairly accurate reliability estimator.

Measuring The Consistency On A Multiple Sequence Alignment

T-Coffee is a heuristic algorithm that attempts to optimize the consistency between a multiple alignment and a list of pre-computed pair-wise alignments known as a library (Figure 2). By consistency we mean that a pair of residues described aligned in the library will also be found aligned in the multiple alignment. While the theoretical maximum for the consistency is 100%, the score of an optimal alignment will only be equal to the level of self-consistency within the library. Figure 2 shows the example of a library that is not self consistent because it is ambiguous regarding the alignment of some of the residues it contains. Of course, the more ambiguous the library, the less consistency it will yield. For instance, given two residues *q* and *r* taken from two different sequences *S1* and *S2*, one can easily measure the consistency ($\mathrm{CS}(R_q^{S1}, R_r^{S2})$) between the alignment of these two residues and all the other alignments contained in the library by comparing $\mathrm{ES}(R_q^{S1}, R_r^{S2})$, the extended score of the pair *q* and *r*, with the sum of the extended scores of all the other potential pairs that involve *S1* and *S2* and either *r* or *q*.

$$\mathrm{CS}\left(R_q^{S1}, R_r^{S2}\right) = \mathrm{ES}\left(R_q^{S1}, R_r^{S2}\right) / \left\{ \sum_{z=S1} \mathrm{ES}\left(R_z^{S1}, R_r^{S2}\right) + \sum_{z=S2} \mathrm{ES}\left(R_q^{S1}, R_z^{S2}\right) \right\} \quad (1)$$

If we want to use it as a quality factor, this measure suffers from two major drawbacks. Firstly it is expensive to compute: given a multiple alignment of N sequences and of length L, each pair of residues found in the multiple alignment needs O(L) operations of extension that require a minimum of O(N) operations each. "O(L)" is a standard notation called *big-O notation*, meaning that the computation time is proportional to L, up to a constant factor. Since there are LN^2 pairs of residues in a multiple alignment, this leads to $O(L^2N^3)$ operations for an estimate of the CS of every pair. This cubic complexity becomes problematic with large numbers of sequences. The second limitation of this measure is that with sequences rich in repeats, the summation factor can become artificially high and cause a dramatic decrease of the consistency score. In practice, we found it much more effective to use the extended score of the best scoring pair contained in the alignment as a normalization factor. This defines the aCS (approximate Consistency measure).

$$\mathrm{aCS}\left(R_q^{S1}, R_r^{S2}\right) = \mathrm{ES}\left(R_q^{S1}, R_r^{S2}\right) / Max\left\{ES\left(R_m^{Sx}, R_n^{Sy}\right)\right\} \quad (2)$$

with R_m^{Sx}, R_n^{Sy} any two residues found aligned in the multiple alignment. Our measurements on the BaliBase dataset indicate that the CS and the aCS are well correlated.

An important criteria, when using the aCS as a reliability measure, is its ability to discriminate between correct and incorrect alignments within the so-called twilight zone (Sander and Schneider, 1991). Given two sequences, the twilight zone is a range of percent identity (between 0 and 30%) that has been shown to be non-informative regarding the phylogenetic relationship that exist among two sequences. Two sequences whose alignment yields less than 30% identity can either be (i) unrelated, (ii) related and incorrectly aligned, or (iii) related and perfectly aligned. To check how good the aCS is when used as an accuracy measure, every 142 BaliBase dataset was aligned using T-Coffee 1.29 and the similarity of each pair of sequence was measured within the obtained alignments. Pairs of sequences with less than 30% identity (5088) were extracted and the accuracy of their alignment was assessed by comparison with their counterparts in the reference BaliBase alignment; the aCS score was also assessed on each pair of aligned residues and averaged along the sequences. Figure 3a shows the scattered graph Identity Vs Accuracy (See Figure legend for the definitions). Despite a weak correlation between these two measurements, the percent identity is a poor predictor of the alignment accuracy. For 75% of the sequence pairs (identity lower than 25%) the accuracy indication given by the percent identity falls in a 40% range (i.e. the average identity indicates the average accuracy +/- 20%). On the other hand, when the accuracy is plotted against the aCS score (Figure 3b) the correlation is improved and for pairs of sequences having an aCS higher than 20 (this is true for 60% of the 5088 pairs) this measure is a much better alignment accuracy predictor than the percent identity. While they do not solve the twilight zone problem, these results indicate that the aCS measure provides us with a powerful mean of assessing an alignment reliability within the twilight zone. Nonetheless, from a practical point of

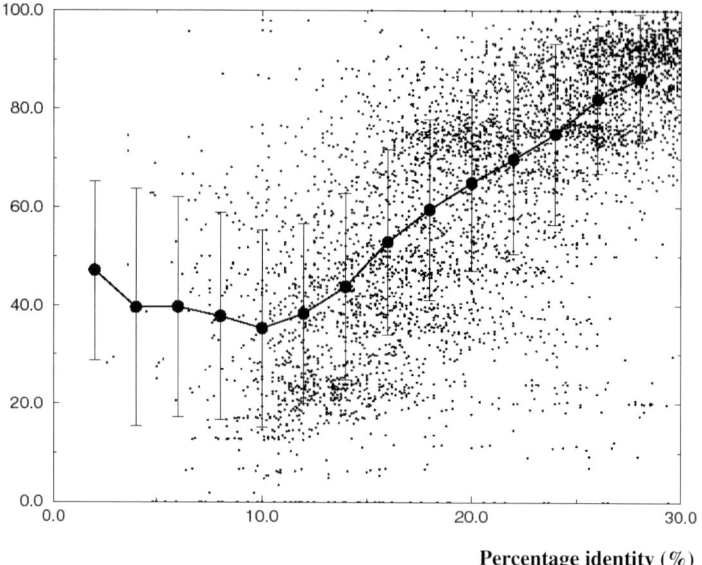

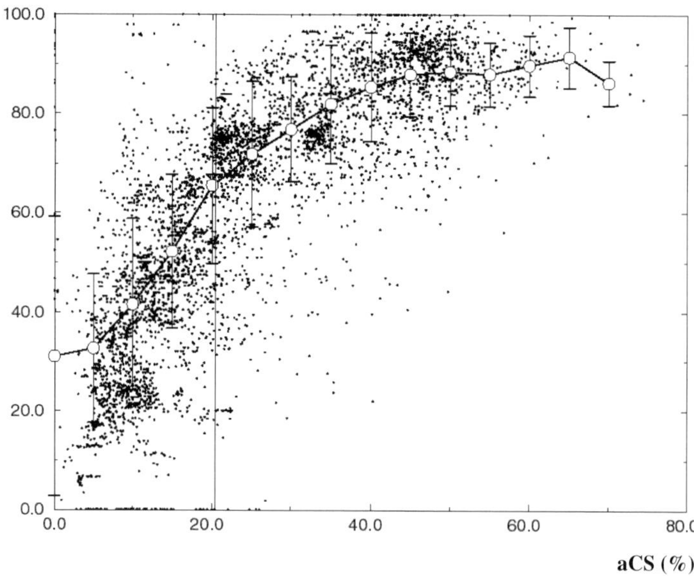

Figure 3. a) Percentage identity Vs Accuracy in the twilight zone: the 5088 pairs of sequences that have less than 30% identity in the BaliBase reference alignments were extracted. The accuracy of their alignment was measured by comparison with the reference, and the resulting graph was plotted. **b)** Approximate Consistency Score (aCS) Vs Accuracy in the twilight zone: the aCS was measured on the 5088 pairs of sequences previously considered and was plotted against the average accuracy previously reported. The vertical line indicates aCS=25 and separates the pairs for which the aCS is informative from those whose aCS seems to be non-informative.

view, the aCS measure is not so useful since one is often more concerned by the overall quality (i.e. is residue r of sequence S correctly aligned to the rest of the sequences?) than by pair-wise relationships. In order to answer this type of questions, the aCS measure was used to derive three very useful non pair-wise indexes.

The Consistency of the Overall Residue Evaluation (CORE)

The CORE is obtained by averaging the scores of each of the aligned pairs involving a residue within a column.

$$\mathrm{CORE}\left(R_q^{Sx}\right) = \sum_{y=1,\, y \neq x}^{N} aCS\left(R_q^{Sx}, R_r^{Sy}\right) / (N-1) \tag{3}$$

where q and r are two residues found aligned in the same column.

The CORE index and equivalent approaches have been shown in the literature to be good indicators of the local quality of a multiple sequence alignment (Heringa, 1999; Notredame *et al.*, 1998), as judged by comparison with reference biological alignments. In the T-Coffee package, an option makes it possible to output multiple alignments with the CORE index (a rounded value between 0 and 9) replacing each residue. It is also possible to produce a colorized version (pdf, postscript or html) of that same multiple alignment where residues receive a background coloration proportional to their CORE index (blue/green for low scoring residues and orange/red for the high scoring ones). Such an output is shown on Figures 4 and 5.

The CORE index described in equation (3) is merely an average aCS measure, and whether that measure provides some indication on the multiple alignment quality is a key question. We tested that hypothesis on the complete BaliBase dataset. Given each T-Coffee alignment, residues were divided in 4 categories: (i) *true positives* (TP) are correctly aligned residues rightfully predicted to be so, (ii) *true negative* (TN) are incorrectly aligned residues rightfully predicted to be so, (iii) *false positive* (FP) are residues predicted to be well aligned when they are not, (iv) *false negative* (FN) are residues wrongly predicted to be misaligned. Following previously described definitions (Notredame *et al.*, 1998), a residue is said to be correctly aligned if at least 50% of the residues to which it was aligned in the reference alignment are found in the same column in the T-Coffee alignment. Each of the 10 CORE indexes (0 to 9) was used in turn as threshold to discriminate correctly and non-correctly aligned residues on the T-Coffee alignments. The BaliBase reference alignments were then used to evaluate the TP, TN,

FP and FN. Sensitivity and the specificity were then computed according to Sneath and Sokal (Sneath and Sokal, 1973) and plotted on a graph (Figure 6). Our results indicate that the best trade off between sensitivity and specificity is obtained when CORE=3 is used as a threshold (i.e. every residue with a score higher or equal to 3 is considered to be properly aligned). In that case the specificity is 84% and the sensitivity is 82%. These high figures partly reflect the overall quality of the T-Coffee alignments in which 80.5% of the residues are correctly aligned according to the criteria used here. It is therefore more interesting to note that when the CORE index reaches 7, the specificity is 97.7% and the sensitivity is close to 50%. This means that thanks to the CORE index, half of the residues properly aligned in a multiple alignment can unambiguously be identified (e.g. more than 40 % of all the residues contained in BaliBase). In the next section we will see that this proper identification sometimes occurs in cases that are far from being trivial, even for an expert eye. Similar results were observed when applying the CORE index on multiple alignments obtained using another method (i.e. ClustalW alignments evaluated with a standard T-Coffee library). This suggests that the CORE measure may be used to evaluate the local quality of a multiple alignment produced by any source. However, one should be well aware that the relevance of the CORE measure regarding the reliability of an alignment is entirely dependant on the way in which the library was derived. All the conclusions drawn here only apply to libraries derived using the standard T-Coffee protocol.

The Sequence CORE (sCORE)

The sCORE is obtained by averaging the CORE scores over all the residues contained within one sequence.

$$\mathrm{sCORE}(Sx) = \sum_{q=1}^{L} (CORE(R_q^{Sx}))/L \tag{4}$$

That measure can be helpful for identifying among the sequences an outlier,

Figure 4. Identifying correct blocks with the CORE measure. An example of the T-Coffee output on a BaliBase test set (1pamA_ref1) that contains five alpha amylases. This alignment was produced using T-Coffee 1.29 with default parameters and requesting the score_pdf output option. The color scale goes from blue (CORE=0, bad) to red (CORE=9, good). The residues in capital are correctly aligned (as judged by comparison with the BaliBase reference). Those in lower case are improperly aligned. Box I indicates a conserved position that, although completely conserved, is not properly aligned. Box II indicates a block of distantly related segments that is correctly aligned by T-Coffee.

```
1pamA  TDVIYQIFTDRFSDGnpannptgaafdgsctnlrlYCGGDWQGIINkinDGYLT
1jdc   --------------------gdciilqgfhwnvvreapnDWYNILR-qqAATIA
1vjs   -----------------lngtlmqyfewympn----DGQHWKRL--QndSAYLA
1bf2   ---------------dviycvhvrgftegdtsipaQYRGTYYGA--GlkASYLA
1smd   -----------------grtsivhlfewrwvdial-----------ecERYLA
                         :       .                    ::

                                    I
1pamA  GMGITAIWISQPVENIYsvinysgvnntAYHGYwa-----------rdfkktNP-
1jdc   ADGFSAIWMPVPWRDFSsws--dgsksgGGEGYfw----------hdfnkn---
1vjs   EHGITAVWIPPA----ykgtsqadvgygaydlydl----------gefhqkgtV
1bf2   SLGVTAVefl-pvqetqndandvvpnsdanqnywgYMTENYFSPDrryaynKA-
1smd   PKGFGGVQVSPPNENVA----ihnpfrpWWERYqp----------vsykl----
        *. .: .  .                 .  *                :

              II
1pamA  --AYGTMQDFKNLIDTAHAHNIKVIIDFAPNHTSPA---ssddpsfae------
1jdc   -GRYGSDAQLRQAASALGGAGVKVLYDVVPNHMNRGypdkein----------
1vjs   RTKYGTKGELQSAIKSLHSRDINVYGDVVINHKGG--adatedvtave------
1bf2   --AGGPTAEFQAMVQAFHNAGIKVYMDVVYNHTAEGGtwtssdpttat------
1smd   CTRSGNEDEFRNMVTRCNNVGVRIYVDAVINHMCGNAvsagtsstcgsYFNPGS
           *   :::          .:.:  * . **            .

1pamA  --------ngrlydngnll--ggytn-----------dtqnlfhhygg-tdfst
1jdc   ------lpagqgfwrndcaDPgnypn-----------dcddgdrfigG------
1vjs   ---vdpadrnrvisgehli--kawthFHFPGRGSTYSdfkwhwyhfd-gtdwde
1bf2   ------iyswrgldnatyy---elts-----------gnqyfydntgIganfnt
1smd   RDFpavpysgwdfndgkck--tg-sg-----------dienyndat--------
                          .                   . .

1pamA  ie--ngiyk---nlydl---ADLNHNNSSVDVYLKDAIKMWL-DLGVDGIRVDA
1jdc   -------------------DADLNTGHPQVYGMFRDEFTNLRSQYGAGGFRFDF
1vjs   srKLnriykFqgkaydYLMYADIDYDHPDVAAEIKRWGTWYANELQLDGFRLDA
1bf2   yn--tvaqn---l--------------------IVDSLAYWANTMGVDGFRFDL
1smd   ----------qvrdcrLSGLLDLALGKDYVRSKIAEYM-nhliDIGVAGFRIDA
                                          :              *:*.*

1pamA  VKHMPFGWQKSFMATINNykpvfT
1jdc   VRGYAPERVNSWMTDSad------
1vjs   VKHIKFSFLRDWVNHVRE------
1bf2   ASVLGnsclngaytasapncpng-
1smd   SKHMWPGDIKAILDKLHN------
               .
```

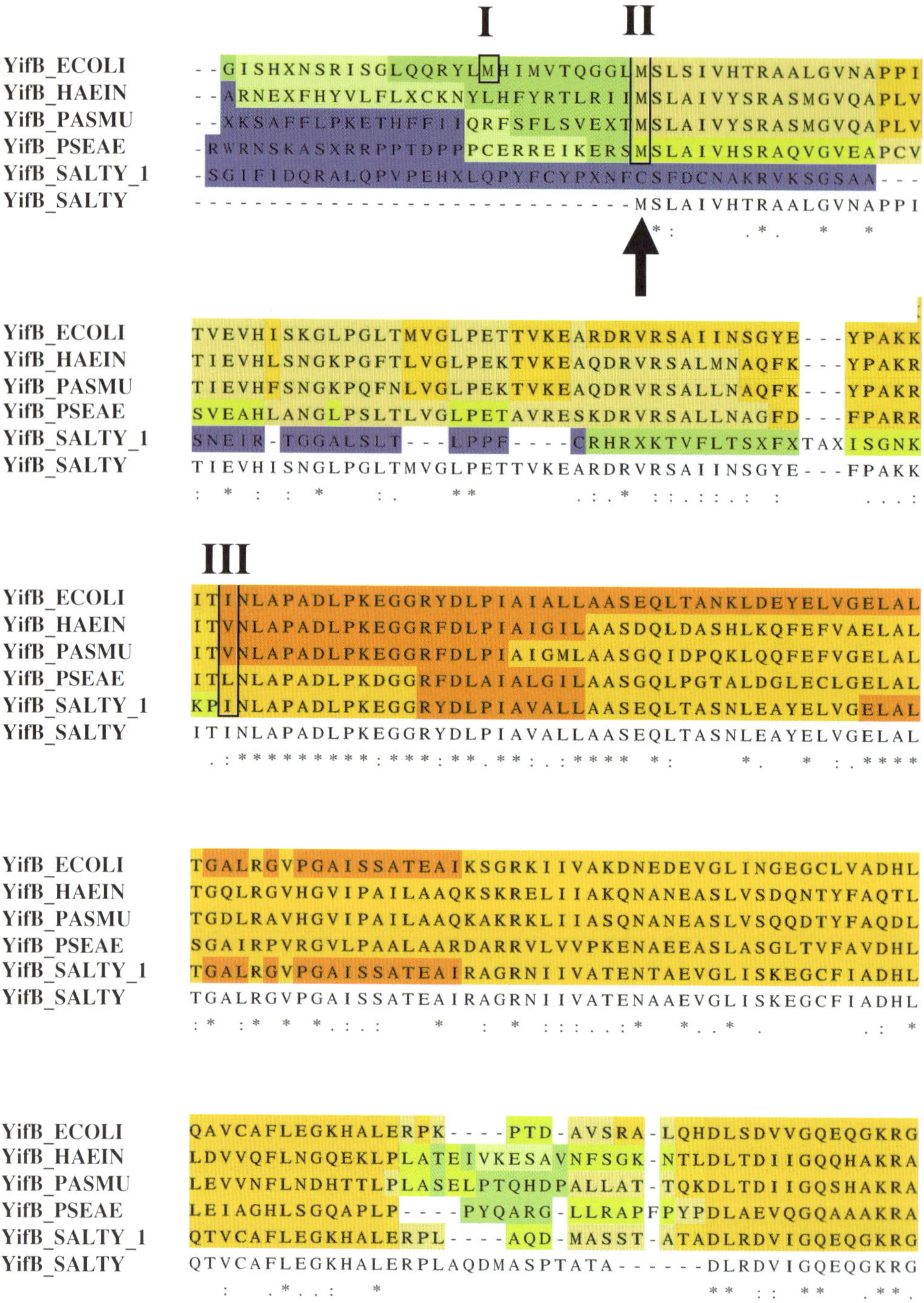

```
                                 I         II
YifB_ECOLI      --GISHXNSRISGLQQRYLMHIMVTQGGLMSLSIVHTRAALGVNAPPI
YifB_HAEIN      --ARNEXFHYVLFLXCKNYLHFYRTLRIIMSLAIVYSRASMGVQAPLV
YifB_PASMU      --XKSAFFLPKETHFFIIQRFSFLSVEXTMSLAIVYSRASMGVQAPLV
YifB_PSEAE      -RWRNSKASXRRPPTDPPPCERREIKERSMSLAIVHSRAQVGVEAPCV
YifB_SALTY_1    -SGIFIDQRALQPVPEHXLQPYFCYPXNFCSFDCNAKRVKSGSAA---
YifB_SALTY      -----------------------------MSLAIVHTRAALGVNAPPI
                              *:     .*.   *  *

YifB_ECOLI      TVEVHISKGLPGLTMVGLPETTVKEARDRVRSAIINSGYE---YPAKK
YifB_HAEIN      TIEVHLSNGKPGFTLVGLPEKTVKEAQDRVRSALMNAQFK---YPAKR
YifB_PASMU      TIEVHFSNGKPQFNLVGLPEKTVKEAQDRVRSALLNAQFK---YPAKR
YifB_PSEAE      SVEAHLANGLPSLTLVGLPETAVRESKDRVRSALLNAGFD---FPARR
YifB_SALTY_1    SNEIR-TGGALSLT---LPPF----CRHRXKTVFLTSXFXTAXISGNK
YifB_SALTY      TIEVHISNGLPGLTMVGLPETTVKEARDRVRSAIINSGYE---FPAKK
                : * : : *    :.   **       .:.* ::.::.: :    ...:

                III
YifB_ECOLI      ITINLAPADLPKEGGRYDLPIAIALLAASEQLTANKLDEYELVGELAL
YifB_HAEIN      ITVNLAPADLPKEGGRFDLPIAIGILAASDQLDASHLKQFEFVAELAL
YifB_PASMU      ITVNLAPADLPKEGGRFDLPIAIGMLAASGQIDPQKLQQFEFVGELAL
YifB_PSEAE      ITLNLAPADLPKDGGRFDLAIALGILAASGQLPGTALDGLECLGELAL
YifB_SALTY_1    KPINLAPADLPKEGGRYDLPIAVALLAASEQLTASNLEAYELVGELAL
YifB_SALTY      ITINLAPADLPKEGGRYDLPIAVALLAASEQLTASNLEAYELVGELAL
                 .:*********:***:**.**:.:****  *:       *.   *  :.****

YifB_ECOLI      TGALRGVPGAISSATEAIKSGRKIIVAKDNEDEVGLINGEGCLVADHL
YifB_HAEIN      TGQLRGVHGVIPAILAAQKSKRELIIAKQNANEASLVSDQNTYFAQTL
YifB_PASMU      TGDLRAVHGVIPAILAAQKAKRKLIIASQNANEASLVSQQDTYFAQDL
YifB_PSEAE      SGAIRPVRGVLPAALAARDARRVLVVPKENAEEASLASGLTVFAVDHL
YifB_SALTY_1    TGALRGVPGAISSATEAIRAGRNIIVATENTAEVGLISKEGCFIADHL
YifB_SALTY      TGALRGVPGAISSATEAIRAGRNIIVATENAAEVGLISKEGCFIADHL
                :*  :*  *  *.:.:    *   :  *  ::::..:*   *..*  .     .: *

YifB_ECOLI      QAVCAFLEGKHALERPK----PTD-AVSRA-LQHDLSDVVGQEQGKRG
YifB_HAEIN      LDVVQFLNGQEKLPLATEIVKESAVNFSGK-NTLDLTDIIGQQHAKRA
YifB_PASMU      LEVVNFLNDHTTLPLASELPTQHDPALLAT-TQKDLTDIIGQSHAKRA
YifB_PSEAE      LEIAGHLSGQAPLP----PYQARG-LLRAPFPYPDLAEVQGQAAAKRA
YifB_SALTY_1    QTVCAFLEGKHALERPL----AQD-MASST-ATADLRDVIGQEQGKRG
YifB_SALTY      QTVCAFLEGKHALERPLAQDMASPTATA------DLRDVIGQEQGKRG
                  :   .*..:   *                     ** ::  **  .**.
```

Figure 5. Identifying frame shifts and start codons. The chosen sequences came are YifB_ECOLI *(Escherichia coli*, accession # AE005174), YifB_SALTY_1 (*Salmonella tiphy*, C18 chromosome, Sanger Center), YifB_HAIN (*Haemophilus influenzae,* Accession # L42023), YifB_PASMU (*Pasteurella multocida*, Accession # AE004439) and YifB_PSEAE (*Pseudomonas aeruginosa*, Accession # AE004091), they were aligned using the standard T-Coffee alignment procedure and requesting the score_pdf output option. The corrected sequence of *Salmonella tiphy* YifB protein sequence was later added for further reference (YifB_SALTY, SP: P57015) but it was not used for coloring the residues (Non colored sequence) or improving the multiple alignment. The figure only shows the N-terminal portion of the alignment, and the arrow indicates the positions annotated as starting codons in SwissProt (except for salmonella tiphy). Box I indicates a putative starting codon in YifB_ECOLI, Box II indicates the true starting codon in most sequences, and Box III indicates the position where the frame-shift occurs in YifB_SALTY_1.

a sequence that should not be part of the set either because it is not homologous or because it is too distantly related to the other members to yield an informative alignment.

The Alignment CORE (alCORE)

The alCORE may be obtained by averaging the sCOREs over all the sequences. Our analysis suggest that this index may be a reasonable indicator of the alignment overall accuracy. Yet, to be fully informative, it requires the sequence set to be homogenous (i.e. the standard deviation of the sCOREs should be as low as possible).

Using The CORE Measure To Assess Local Alignment Quality

The driving force behind the development of the CORE index is the identification of correctly aligned blocks of residues within a multiple sequence alignment. It is common practice to identify these blocks by scanning the multiple alignment and marking highly conserved regions as potentially meaningful. ClustalW and ClustalX provide a measure of conservation that may help the user when carrying out this task. Unfortunately, situations exist where it is difficult to make a decision regarding the correct alignment of some residues within an alignment. Such an example is provided in Figure 4 with the BaliBase alignment known as 1pamA_ref1, made of six alpha-amylases.

This set is difficult to align because it contains highly divergent sequences. Not only have these sequences accumulated mutations while they were diverging but they have also undergone many insertion/deletions that make

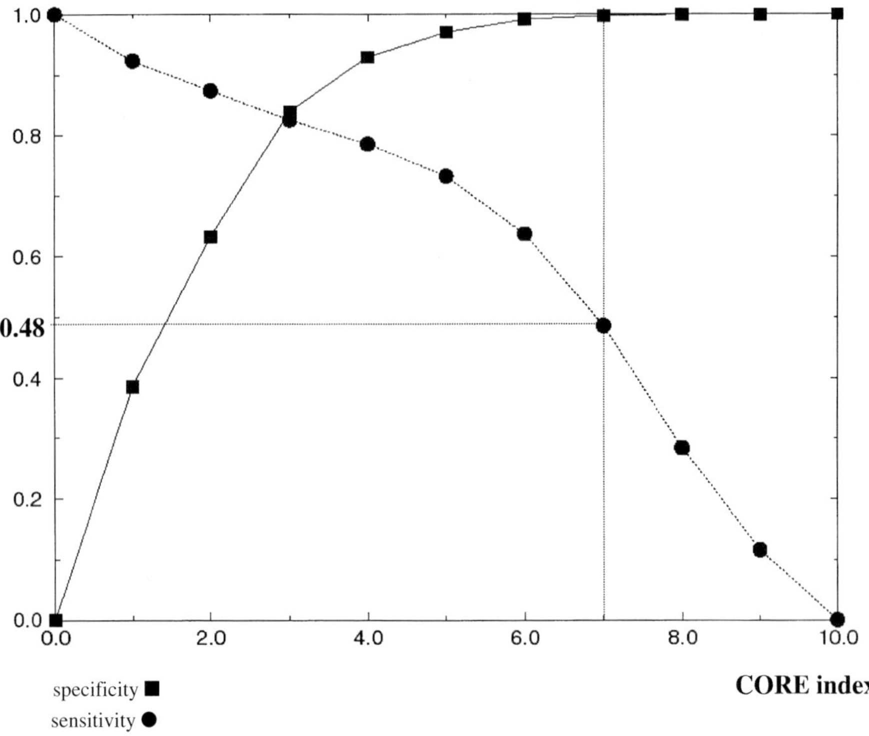

Figure 6. Specificity and sensitivity of the CORE measure. The sensitivity and the specificity of the CORE index used as an alignment quality predictor were evaluated on the BaliBase test-sets. Measures were carried out on the entire BaliBase dataset. The sensitivity (●) and the specificity (■) were measured on the T-Coffee alignments after considering that every residue with a CORE index higher than x was properly aligned (see text for details).

it difficult to reconstruct their relationships with accuracy. The average level of identity measured on the BaliBase reference is 18%, the two closest sequences being less than 20% identical. As such, 1pamA_ref1 constitutes a classic example of a test set deceptive for most multiple sequence alignment methods. The fact that less than one third of the 1pam_ref1 reference alignment is annotated as trustable in BaliBase confirms that suspicion. When ran on this test-set, existing alignment programs generate different results, Prrp finds 37% of the columns annotated as trustable in BaliBase, ClustalW (1.81) 40%, T-Coffee 54% and Dialign2 56%. Regardless of the methods used, such an alignment is completely useless unless correctly aligned portions can be identified. It is exactly the information that the CORE index provides us with. An alignment colorized according to its CORE indexes is shown on Figure 4.

The results are in good agreement with those reported in Figure 6. Out of the 905 correctly aligned residues (42% of the total), 267 have a score higher than 7. No incorrectly aligned residue has a score higher or equal to 7. Using 7 as a prediction threshold gives a sensitivity of 29% and a specificity of 100%. Residues with a CORE index of 3 or higher (yellow pale) yield a sensitivity of 65% and a specificity of 79%. In this alignment, the main features are the red/dark-orange blocks: they are 100% correct. These blocks could be fed as they are to any suitable method (structure prediction, phylogeny....). They are not very well conserved at the sequence level and are therefore very informative for structural and phylogenetic analysis. For instance, the block II in Figure 4 is perfectly aligned although within that block, the average pair-wise identity is lower than 30% (41% for the two most closely related sequences). The measure of consistency can also help questioning positions that may seem unambiguous from a sequence point of view. In the column annotated as I, the position marked with a "*" could easily be mistaken to be correct: it is within a block, aromatic positions are usually fairly well conserved and owing to their relative rarity, unlikely to occur by chance. Yet the green color code indicates that this position may be incorrectly aligned (the green tyrosine has a CORE index of 1). This is confirmed by comparison with the reference that shows the correct alignment to incorporate another tyrosine at this position.

When analyzing these patterns, one should always keep in mind that the consistency information only has a positive value. In other words, inconsistent regions are those where the library does not support the alignment. This does not mean they are incorrectly aligned but rather that no information is at hand to support or disprove the observed alignment.

Identifying Faulty Gene Predictions

Another possible application of the T-Coffee CORE index is to reveal and help resolving sequence ambiguities in predicted genes. In the structural genomic era, many projects involve hypothetical proteins, for which an accurate prediction of the start and stop codon is needed to properly express the gene product. Since over-predicted N or C-terminal are rarely conserved at the amino acid level, sequence comparison provides us with a very powerful mean of identifying this type of problems. A simple procedure consists of multiply aligning the most conserved members of a protein family before measuring the T-Coffee CORE index on the resulting alignment. Inspection of the CORE patterns offers a diagnostic regarding the correctness of the data. This approach can also be applied to frame-shift detection where the identification of abnormally low scoring segments may lead to their correction. Such an alignment will make it possible to decide if the abnormal length of a coding region could be due to a sequencing error (and the resulting frame-shift). At least the CORE measure will indicate that a thorough examination is needed. Of course, one could also detect these frame-shifts using standard pair-wise comparison methods such as Gene-wise (Birney and Durbin, 2000), but the advantage of using a multiple sequence alignment is that the simultaneous comparison of several sequences can strengthen the evidence that the frame-shift is real. Furthermore, thanks to the multiple alignment, one may be able to detect mistakes in sequences that lack a very close homologue.

To illustrate this potential usage of T-Coffee, we chose the example of an *Escherichia coli K12* gene (Accession # U00096) predicted to encode a protein of unknown function, yifB. Orthologous genes were found in complete genomes using BLAST (Altschul *et al.*, 1997) and the four most conserved sequences (identity >70% relative to the *Escherichia coli K12* gene, see figure for ID numbers) were retrieved along with their flanking regions (80 nucleotides on the N-terminus side) in order to check whether these supposedly non coding regions did not contain any coding information. The 'elongated' sequences were translated in the same frame as their core coding region, their multiple alignment was carried out using T-Coffee, and the CORE indexes were measured. The resulting alignment is displayed on Figure 5 with the CORE indexes color-coded (low CORE in blue and green, high CORE in orange and red). The main feature on the N-terminus is an abrupt transition (II) from low to high CORE indexes. This position is also a conserved methionine. The combination of these two observations suggests that the starting point of these five sequences is probably where the transition occurs, ruling out other methionines as potential starting points in the first sequence (I). Another discrepancy occurs in this alignment that is also emphasized by the CORE analysis: the sequence yifB_SALTY_1 yields a very low N-terminal CORE index, relatively to the other family members. The CORE score of this sequence becomes abruptly in phase with the other sequences at the position marked III. This pattern is a clear indication of a frame-shift: a protein highly similar to the other members of its family but locally unrelated. To verify that hypothesis, we used some data provided by SwissProt (Bairoch and Boeckmann, 1992) and found that in the corresponding entry, the nucleotide sequence has been corrected to remove the frame-shift we observed (entry P57015). The corrected sequence has been added to the bottom of the alignment on Figure 5 (non-colored sequence). The position where yifB_SALTY_1 and its corrected version start agreeing is also the position where the CORE score changes abruptly from a value of 2 (yellow) to a value of 7 (orange). That position also turns out to be the one where the frame-shift occurs in the genomic sequence.

Conclusion

In this chapter, we introduced an extension of the T-Coffee multiple sequence alignment method: the CORE index. The CORE index is a mean of assessing the local reliability of a multiple sequence alignment. Using the CORE index, correct blocks within a multiple sequence alignment can be identified. This measure also makes it possible to detect potential errors in genomic data, and to correct them. The CORE index is a relatively ad hoc measure and even if it may prove extremely useful from a practical point of view, it still needs to be attached to a more theoretical framework. One would really need to be able to turn the consistency estimation into some sort of P-Value. For instance, to assess efficiently the local value of an alignment, one would like to ask questions of the following kind: what is the probability that library X was generated using dataset Y? What is the probability that alignment A yields p% consistency with library X? Altogether these questions may open more venues to the automatic processing of multiple alignments. That issue may prove crucial for the maintenance of resources that rely on a large scale usage of multiple sequence alignments such as Hobacgene (Perriere *et al.*, 2000), Hovergene (Duret *et al.*, 1994) or ProDom (Corpet *et al.*, 2000).

References

Altschul, S.F., Madden, T.L., Schaffer, A.A., Zhang, J., Zhang, Z., Miller, W. and Lipman, D. 1997. Gapped BLAST and PSI-BLAST: a new generation of protein database search programs. nucleic acids res. 25: 2289-3402.

Bairoch, A. and Boeckmann, B. 1992. The SWISS-PROT protein sequence data bank. Nucleic Acids Res: 2019-2022.

Birney, E. and Durbin, R. 2000. Using GeneWise in the Drosophila annotation experiment. Genome Res 10: 547-548.

Corpet, F., Servant, F., Gouzy, J. and Kahn, D. 2000. ProDom and ProDom-CG: tools for protein domain analysis and whole genome comparisons. Nucleic Acids Res. 28: 267-269.

Dayhoff, M.O., Schwarz, R.M. and Orcutt, B.C. 1979. A model of evolutionary change in proteins. Detecting distant relationships: computer methods and results. In: Atlas of Protein Sequence and Structure. M.O. Dayhoff, editor. National Biomedical Research Foundation, Washington, D.C. p. 353-358.

Duret, L. and Abdeddaim, S. 2000. Multiple alignment for structural, functional, or phylogenetic analyses of homologous sequences. In: Bioinformatics, Sequence, Structure and Databanks. D. Higgins and W. Taylor, eds. Oxford University Press, Oxford.

Duret, L., Mouchiroud, D. and Gouy, M. 1994. HOVERGEN: a database of homologous vertebrate genes. Nucleic Acids Res. 22: 2360-2365.

Feng, D.-F. and Doolittle, R.F. 1987. Progressive sequence alignment as a prerequisite to correct phylogenetic trees. J. Mol. Evol. 25: 351-360.

Gotoh, O. 1996. Significant improvement in accuracy of multiple protein sequence alignments by iterative refinements as assessed by reference to structural alignments. J. Mol. Biol. 264: 823-838.

Gribskov, M., McLachlan, M. and Eisenberg, D. 1987. Profile analysis: Detection of distantly related proteins. Proceedings of the National Academy of Sciences 84: 4355-5358.

Henikoff, S. and Henikoff, J.G. 1992. Amino acid substitution matrices from protein blocks. Proc. Natl. Acad. Sci. 89: 10915-10919.

Heringa, J. 1999. Two strategies for sequence comparison: profile-preprocessed and secondary structure-induced multiple alignment. Computers and Chemistry 23: 341-364.

Huang, X. and Miller, W. 1991. A time-efficient, linear-space local similarity algorithm. Adv. Appl. Math. 12: 337-357.

Jones, D.T. 1999. Protein secondary structure prediction based on position-specific scoring matrices. J. Mol. Biol. 292: 195-202.

Krogh, A., Brown, M., Mian, I.S., Sjölander, K. and Haussler, D. 1994. Hidden Markov models in computational biology: Applications to Protein Modeling. J. Mol. Biol. 235: 1501-1531.

Morgenstern, B., Frech, K., Dress, A. and Werner, T. 1998. DIALIGN: finding local similarities by multiple sequence alignment. Bioinformatics 14: 290-294.

Needleman, S.B. and Wunsch, C.D. 1970. A general method applicable to the search for similarities in the amino acid sequence of two proteins. J. Mol. Biol. 48: 443-453.

Notredame, C., Higgins, D.G. and Heringa, J. 2000. T-Coffee: A novel algorithm for multiple sequence alignment. J. Mol. Biol. 302: 205-217.

Notredame, C., Holm, L. and Higgins, D.G. 1998. COFFEE: an objective function for multiple sequence alignments. Bioinformatics 14: 407-422.

Perriere, G., Duret, L. and Gouy, M. 2000. HOBACGEN: database system for comparative genomics in bacteria. Genome Research 10: 379-385.

Saitou, N. and Nei, M. 1987. The neighbor-joining method: a new method for reconstructing phylogenetic trees. Mol. Biol. Evol. 4: 406-425.

Sander, C. and Schneider, R. 1991. Database of homology-derived structures and the structurally meaning of sequence alignment. Proteins: Structure, Function, and Genetics 9: 56-68.

Sneath, P.H.A. and Sokal, R.R. 1973. Numerical Taxonomy. Freeman, W.H., San Francisco.

Taylor, W.R. 1988. A flexible method to align large numbers of biological sequences. J. Mol. Evol. 28: 161-169.

Thompson, J., Higgins, D. and Gibson, T. 1994. CLUSTAL W: improving the sensitivity of progressive multiple sequence alignment through sequence weighting, position-specific gap penalties and weight matrix choice. Nucleic Acids Res. 22: 4673-4690.

Thompson, J.D., Plewniak, F. and Poch, O. 1999. A comprehensive comparison of multiple sequence alignment programs. Nucleic Acids Res. 27: 2682-2690.

From: *Bioinformatics and Genomes: Current Perspectives*
Edited by: Miguel A. Andrade

Chapter 3

Analysis of Expression Data

Javier Tamames

Abstract

DNA array technology has become the most powerful tool in functional genomics. It is capable of monitoring simultaneously the expression profiles of thousand of genes in different conditions, hence it is an invaluable tool for exploring the responses of an organism. In addition, the potential of array techniques for exploring mutations and polymorphisms in genes is enormous.

From the bioinformatics point of view, a microarray experiment involves high complexity, both in the preparation and management of the experiment, and in the processing of the results. The difficulties are summarized as follows: To prepare microarrays, high numbers of clones are used, with the expectations that arrays containing hundred of thousand of genes will be soon available. The management of the collections of clones must be planned carefully, and every clone must be tracked in terms of its location, function and performance in the experiments. In the management of the experiment, the creation of interfaces to control the machinery involved in the experiment is desirable but difficult, since different vendors exist and no standard is being used. After the hybridisation step, accurate imaging software must be used in order to obtain clear results.

The analysis of the results obtained can be performed using different approaches: Clustering algorithms are widely used, however other methods have been used as well. It is important to know the advantages and drawbacks of the different methods in order to obtain meaningful information. Once groups of related genes have been found, a very important and often ignored element is the data-mining on the results, to extend the biological information about such groups of genes.

Finally, it is very important to store all the information relative to the experiment, using relational schemes of storage. To date, even though several different schemes have been proposed, no single standard exists. This makes it difficult to exchange and compare experiments between laboratories. In this chapter, I address in detail some of these points and propose possible solutions.

Introduction

Various technologies using the similar principles of hybridisation and recognition of specific DNA sequences are grouped under the term DNA matrices. The most powerful and commonly-used technique is that involving microarrays, due to its capacity for the study of large groups of genes. This technique allows the simultaneous monitoring of the expression of thousands of genes, together with the comparison of expression patterns in varying cellular states.

Microarrays have been used for studying a wide range of different issues. For genomics and proteomics, DNA arrays have been used for discovering regulatory elements for genes that display common patterns of gene expression (Cohen *et al.*, 2000; Leemans *et al.*, 2001); protein function prediction for related genes (Brown *et al.*, 2000; Cummings and Relman, 2000); genotyping (Behr *et al.*, 1999); and genomic analysis of non-sequenced organisms (Akman and Aksoy, 2001; Hayward *et al.*, 2000).

In relation to metabolomics, DNA arrays are useful tools for studying the organization of biological processes. Analyses for genetic networks, metabolic pathways and the temporal development of biological processes have been carried out for a diverse set of organisms (including humans) (deRisi *et al.*, 1997; Ferea and Brown, 1999; Ideker *et al.*, 2001; Iyer *et al.*, 1999; Richmond *et al.*, 1999; Tavazoie *et al.*, 1999).

Also on the clinical side, microarrays are finding very valuable applications. Drug discovery is a field that particularly benefits from the use of DNA array technologies (Debouck and Goodfellow, 1999). These have been successfully applied to drug target identification (Kozian and

Kirschbaum, 1999), development (Gray *et al.*, 1998) and validation (Marton *et al.*, 1998; Wilson *et al.*, 1999). Pharmacogenomics is expected to develop principally via the use of DNA array information (Evans and Relling, 1999; Scherf *et al.*, 2000). Diagnostic research is also expected to benefit, with preliminary and very promising results having been produced for the diagnosis of cancer (Golub *et al.*, 1999; Scherf *et al.*, 2000). Pathogenicity mechanisms and disease progression can also be followed accurately (Cummings and Relman, 2000; Geiss *et al.*, 2000).

There are obvious advantages to being able to rapidly study practically all of an organism's genes. However in order to make the most of this technique it is essential to have access to appropriate bioinformatics tools and equipment, and thereby to be able to store, organise and query the data obtained so as to convert them into results and to draw appropriate conclusions.

Experimental Approaches

From the experimental point of view there are two main approaches to carrying out microarray studies:

1. The user is responsible for the design of the entire experiment. This implies maintaining a collection of clones which can be used to decide which genes are of interest for the study. The array is generated *in situ* using a robotic device, and after hybridisation has taken place, the image resulting from the experiment is taken. This image is then the starting point for analysis of the results. The main advantages of this approach are the flexibility and control it offers (the user decides what is to be placed on the array and at all stages knows how it has been created). The main disadvantages lie in the initial cost and the larger scale of infrastructure required.

2. A pre-fabricated array is bought from a commercial provider. This will contain a collection of oligonucleotides or cDNAs which are representative of certain species, tissues, etc. It is therefore not necessary for the user to maintain a collection of clones or to own equipment for physically creating the array. The commercial providers usually also supply the hardware and software needed for image analysis. For the rest of the process there is no real difference from the previous approach, although some providers may also offer software for analysis of the results, but this is usually somewhat limited. The main advantages of

this approach are its ease of use and lower initial cost. The disadvantages include the limited potential for varying the protocols or analysis tools, and the purchase price of each array.

Under the first approach, carrying out the complete design of the experiment will require bioinformatics tools that allow maintenance of the clone collection and the collation of information on their characteristics, in order to be able to know which are of interest for the study. Bioinformatics tools are also useful for aiding the process of creating the array via interfaces to the robot equipment. Once the image has been obtained and quantified, these approaches then converge. Therefore, the bioinformatics systems needed for microarray technology can be divided into two groups:

- Solutions for the design and carrying out of experiments (first approach).
- Solutions for analysis of the results and storage of the data (both approaches).

Design Of The Experiment

The application of microarray technology requires the manipulation of a large collection of clones. It is essential to have a system in place for monitoring and managing these clones that allows the user to locate each clone on its corresponding plaque and to access the information available for that clone.

It is therefore necessary to create a database that stores this information and that is simple and efficient to use. This database should incorporate various types of information, including:

- Identification of each clone.
- Conditions in which the clone was obtained, characterized, and stored.
- Location of the clone within the storage medium, and the existence of any replicas.
- Data on use of the clone in previous arrays, and the quality controls in place for the results obtained in those experiments.
- Information on the functionality of the clone, its location, the tissues in which it is expressed, and its participation in metabolic pathways, etc.

The last point is especially important: except for certain special cases (Loftus *et al.*, 1999; Rockett *et al.*, 2001), "brute force" approaches are typically used that consist of putting as many genes as possible onto an array. At best some form of pre-screening is carried out using various arrays with all possible genes present and then a selection is made from these *a posteriori*. Whichever solution is chosen in the end, it is clear that pre-selecting genes based on prior knowledge of their function, the tissue in which they are expressed, etc, should always improve the effectiveness of the experiment.

Pre-selection of genes is possible using a system based on a list of genes that includes information from various sources, allowing efficient retrieval of genes with the desired characteristics. This requires data mining systems that refer to both the available literature and other databases and extract and integrate the desired information. In the same way, the system should also allow for automatic searching for new genes with the desired properties that are not present in the clone collection so that new genes of interest may be identified.

This database should be easy to update so that new information can be added (e.g. new clones, new plates, quality controls for new experiments, new functional information, etc). It should also be directly linked to the processing of the experimental data so that the performance of each clone in previous experiments may be monitored. The search engine associated with the database should allow complex queries to be carried out on the clone information, e.g. by combining the data from various different fields. In this way the user can determine which are the most relevant clones for the experimental study and can monitor these.

Interface To Machinery

The array design system will ideally connect with a system for controlling the robotic machinery involved in the experiment. This system should offer the possibility of creating a new plate based on the clones selected by the user from which the array will be generated. These robotic devices are usually controlled by the software supplied with them. Given the variety of existing robots and formats of the input and output data that they use, the software that is created for this stage must be able to be adapted to the various different devices.

Analysis Of The Image

There are several public and commercial solutions for the problem of analysis of the images produced in DNA array experiments. All of them are capable of working with the standard *.TIFF* images obtained in fluorescence-based experiments, and in some cases the *.GEL* images from radioactive macroarray experiments.

The quality of this type of program can be measured according to the effectiveness with which the program solves the main steps in the process of image analysis. These steps are:

- Image capture
- Adjustment of the grid
- Detection of the spots
- Detection and annotation of low quality spots
- Background estimation
- Quantification of the spots

A detailed discussion of these issues is outside the scope of this chapter. A good review can be found in Zhou *et al.* (2000).

Processing And Quality Control

Once the data generated from the array have been quantified, it is then necessary to process them such that points of low quality are disregarded. The information generated from this quality control should ideally be passed to the clone management system in order to allow monitoring of the results of each clone.

The data obtained must then be normalised so that they meet the distribution and variance requirements necessary for the subsequent statistical analysis. There are various different normalisation methods that are more or less appropriate depending on the type of experiment. At this stage a bioinformatics system will offer various alternative options for normalising the data so that the most appropriate method for each problem may be chosen.

Analysis: Clustering Of Genes With Similar Behaviour

The simplest DNA matrix experiment is that where there are just two separate conditions (e.g. trials in the presence and then the absence of a drug). In this type of experiment only two values are obtained for each gene and the data are reasonably easy to interpret (and in fact the image analysis programs that quantify the points of the matrix often incorporate simple utilities for carrying out this task). This may consist of a represention of the expression values of the genes under the two conditions using a line graph, which will indicate which points (genes) increase or reduce their expression (ie. lie off the diagonal).

In case that the experiment is more complex and the number of conditions is increased, this simple type of analysis is no longer valid. This is the case for example when different concentrations of a drug are being tested, or the expression of genes is being monitored at various points over a period of time (time series). In these cases the amount of information to be analysed can run into the tens or hundreds of thousands of data points. This large-scale automatic generation of data with no human intervention can lead to high levels of noise, and can mean that some data are missing due to low levels of quality or experimental errors. Due to the large quantity of information, visual inspection of the data is not feasible and it is necessary to employ a process of classification of the data in order to obtain initial conclusions.

The most common and informative analysis made of the data from a microarray experiment is termed clustering (or grouping). Both genes and conditions can be grouped. Grouping genes allows identification of those genes exhibiting similar expression patterns, whilst grouping by conditions allows identification of those conditions giving rise to similar gene expression. We can imagine the results of an experiment as forming a matrix, with the rows representing genes and the columns representing conditions. Each element of the matrix then represents the expression value for a specific gene under a specific condition, and we can then either group rows (genes) or columns (conditions). These classifications allow us to detect genes that are co-expressed and that therefore may be subject to common transcription control (e.g. via the same promoters, etc) or be implicated in similar cellular functions.

The clustering is usually performed based on a set of distances between all the expression profiles. Pairwise comparison of gene expression profiles must therefore be computed. Various different definitions of distance are commonly used, including correlation coefficients, Euclidean distance, and mutual information. Correlation coefficients are those that are likely to be most robust against the small variations and noise in the experimental data. Once distances have been obtained, the next step is to compare and group the patterns into clusters of genes possessing correlated profiles. Different approaches have been applied to the comparison of large numbers of expression patterns, including hierarchical clustering, multivariate analysis and neural networks (Eisen *et al.*, 1998; Tamayo *et al.*, 1999; Törönen *et al.*, 1999).

Hierarchical clustering (Sneath and Sokal, 1973) has been used in a wide number of cases (Eisen *et al.*, 1998; Iyer *et al.*, 1999), and is the most commonly used way to analyze patterns of gene expression, probably because it is intuitive and easy to understand. The most similar patterns are clustered together in a hierarchy of nested subsets, producing a binary tree. The major drawback of using hierarchical clustering is that expression data usually contain a non-negligible amount of noise, and classical clustering techniques are known to be very sensitive to this situation. Moreover, it has been noted (Sneath and Sokal, 1973) that hierarchical clustering suffers from a lack of robustness and the solution can depend on the order of the data input (Tamayo *et al.*, 1999). Computing time is also a disadvantage, since this increases quadratically as data size increases. Therefore, the analysis of experiments with a large number of genes and conditions (as is currently the case) is greatly impeded.

Neural networks have a series of properties that make them suitable for the analysis of gene expression patterns. They are able to deal with real-world data sets containing noisy, ill-defined items with irrelevant variables and outliers, and whose statistical distributions do not need to be parametric. Therefore, neural networks have been proposed as an alternative to classical hierarchical clustering.

Self-organising maps (SOMs)(Kohonen, 1990) are a type of network that provides a robust approach to the clustering of large amounts of noisy data. The SOM structure is a two-dimensional grid, usually of hexagonal or rectangular geometry, with the number of nodes being fixed from the beginning. A training process is performed in which the profiles are shown to the system and the nodes change according to the distribution of variability of the data set. At the end of the process, similar gene expression patterns map close together in the network. SOMs are reasonably fast and can easily be scaled to large data sets, but they present several problems (Fritzke, 1994): SOMs fix the number of clusters arbitrarily from the beginning, which makes reproduction of the "natural" cluster structure of the data very dependent on the subjective choice of the number of clusters. Additionally, the clustering obtained is not proportional: if any particular type of profile is abundant, the SOM will produce an output in which this type of data populates the vast majority of clusters, whilst other profiles will map into just a few clusters, with reduced resolution. Finally, the lack of a tree structure makes it impossible to detect higher order relationships between clusters of profiles.

A new neural network approach has been proposed recently that combines the advantages of the other methods. The SOTA algorithm (Self-Organising Tree Algorithm; Herrero *et al.*, 2000; http: //www.almabioinfo.com/SOTA) is capable of producing hierarchical tree structures that therefore facilitate the representation of higher order relationships between groups of profiles, without the loss of the advantages of the direct classifications produced via unsupervised learning methods. SOTA is able to deal with expression array data with considerable amounts of noise, and since the execution time of the algorithm is linearly dependent on the number of samples, it is suitable for analyzing large sets of data.

The SOTA structure grows from the root of the tree, where all the

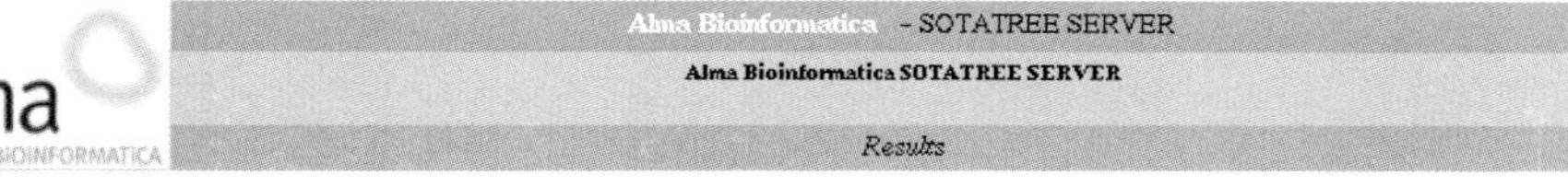

Newick File: yeast_cell_cycle_1936_19422317835361292089063600391917617.nw

Output File: yeast_cell_cycle_1936_19422317835361292089063600391917617.png

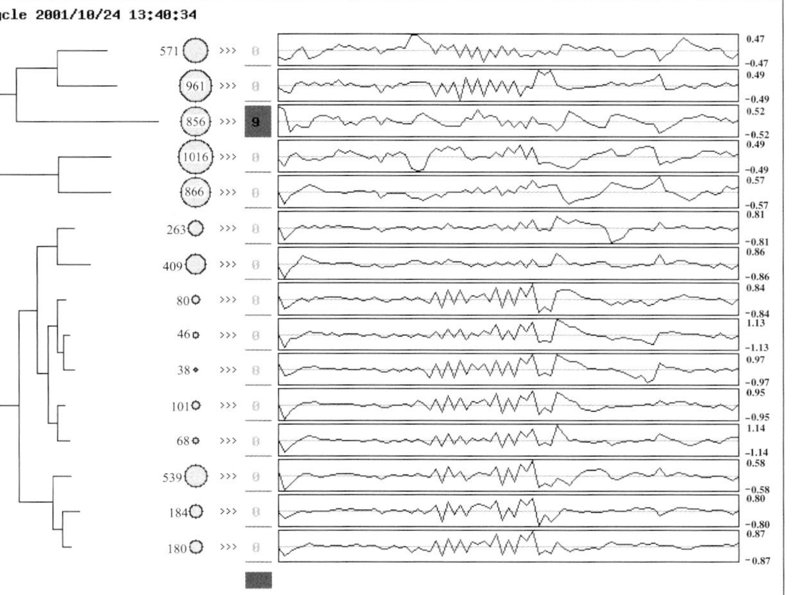

Figure 1. SOTA is an efficient clustering algorithm, capable of quickly analyzing vast amounts of data. The results show an expandible hierarchical tree structure, allowing the user to explore and find the most appropiate level of clustering. (http://www.almabioinfo.com/SOTA).

expression profiles are mapped to one node, towards the leaves which will contain only a single profile (Figure 1). The final result is a binary tree topology that incorporates the principles of the growing cell structures algorithm of Fritzke (1994). This structure can be asymmetrical, including branches with different numbers of nodes. Growth can be stopped at any desired level in order to adjust the level of resolution of the clustering to the particular needs.

There are several features that increase the robustness of this algorithm. An essential point is that every single data item contributes information to the clustering process at all levels. Since clustering is driven by the variability, the number of data items has little influence on the definition of the final cluster structure of the data set. Thus if a given type of profile is abundant, all items similar to this profile will remain grouped together in a single cluster and will not directly affect the rest of the clustering process performed by the network.

The SOTA approach also has additional advantages: the profile values associated with the nodes are equivalent to a weighted average of the corresponding profiles, and can be used directly as representative profiles of the associated genes. The similarity of the expression patterns associated with each node can be used directly to estimate the quality of the clusters. And finally the binary tree representation is appropriate for viewing and representing the data.

Storage Of The Information

Storage of the data is of fundamental importance from the bioinformatics point of view. Easy access is critical, as well as the ability to perform complex queries in a highly-optimised way and to combine the information obtained with that from other experiments. It is therefore essential to develop databases that allow the data to be structured in a rational way. One of the most commonly-used technologies is that of a relational database manager (Postgres SQL, Oracle, etc).

The amount of information generated by a microarray experiment is very high. It is not just the image and clustering results that will need to be stored, but also the conditions under which the experiment was carried out, the characteristics of the sample analysed, the experimental protocols, etc. One of the challenges of microarray technology is to generate data that are easily interpreted and reproducible. It is therefore essential to develop standards for the storage of the information generated by experiments such that all the relevant aspects are documented. For this the MGED (Microarray

Gene Expression Database group, http: //www.mged.org) has produced MIAME (Minimum Information About a Microarray Experiment), a document that specifies the minimum information to be stored for each experiment. Several different database schema are proposed for the storage of these data.

The main DNA array repository projects currently include:

- ArrayExpress (European Bioinformatics Institute (EBI)). http: //www.ebi.ac.uk/arrayexpress/
- GeneX (National Center for Genome Research (NCGR), University of California). http: //www.ncgr.org/genex/
- GEO ("Gene Expression Omnibus"). National Center for Biotechnology Information (NCBI), USA. http: //www.ncbi.nlm.nih.gov/geo/
- SMD ("Stanford Microarray Database"). http: //www.dnachip.org/

All of these systems allow integration of the essential data of a microarray experiment within a relational database, and generally can be implemented at different levels of complexity.

Annotation

By performing an experiment and then analysing the results using clustering methods, we identify groups of genes that respond in the same way to the experimental conditions of the study. It is at this point that the researcher will need to know if the classification so obtained is sensible from a biological point of view and what conclusions may be drawn from the results. Examples of some of the questions that may arise at this point are:

- In which metabolic pathway, cellular process or signalling pathway does a candidate gene take part?
- What alternative functions may a certain protein have in other organisms?
- What members of the family of proteins under study are transcriptionally regulated by a given factor? Is any unrelated mRNA also similarly regulated?
- What do the products of the mRNAs showing very similar expression patterns have in common during the life of the cells of interest?
- Which genes should be included in future experiments? Where can they be found? Which researcher has the most experience of these and what are the most relevant publications?

Independently of whether an experiment has been carefully designed to meet a specific objective, it is common for new questions and research areas to be suggested as the information is generated. The post-analysis stage is therefore of fundamental importance, adding a new dimension of information to the results of the experiment. Two complementary systems for this are information extraction (IE) and ontology-based annotation.

Information Extraction

IE is now starting to be used in the context of microarray experiments. The goal is to gather all relevant information on the genes making up the clusters, and uncover common features that can offer useful clues as to the functionality and biological relevance of the clusters. One example of this type of tool is the ALMATextMiner system (http: //www.almabioinfo.com/ATM) (Figure 2). The system uses a hierarchical classification of genes within clusters, and extracts information from the literature (Medline abstracts) that is relevant to the groups of genes produced in the classification. From this it obtains the characteristics that are common to these genes, together with descriptions of their biological function. In this way keywords common to the genes of a group are identified, and these will usually describe functional aspects of these genes (e.g. phrases of the type "genes responsible for the repair of damage to DNA"). At the same time, given that hierarchical trees produce classifications at different levels, the most discriminating level can be determined, or that which connects various functions together. This greatly facilitates interpretation of the results of the experiment as the researcher can immediately access information on the functionality of the gene groups that have been identified according to their expression patterns. Searches for related genes not present in the experiment can also be performed, based on the information in the literature.

Ontologies

Classification of functions into hierarchical schema of functional classes allows better description and understanding of the features of biological entities such as genes or proteins. It also provides a unified framework for assessing the relationships between such entities in terms of function, and a direct way for comparing their functional annotations and inferring the possible biological meaning of groupings of them.

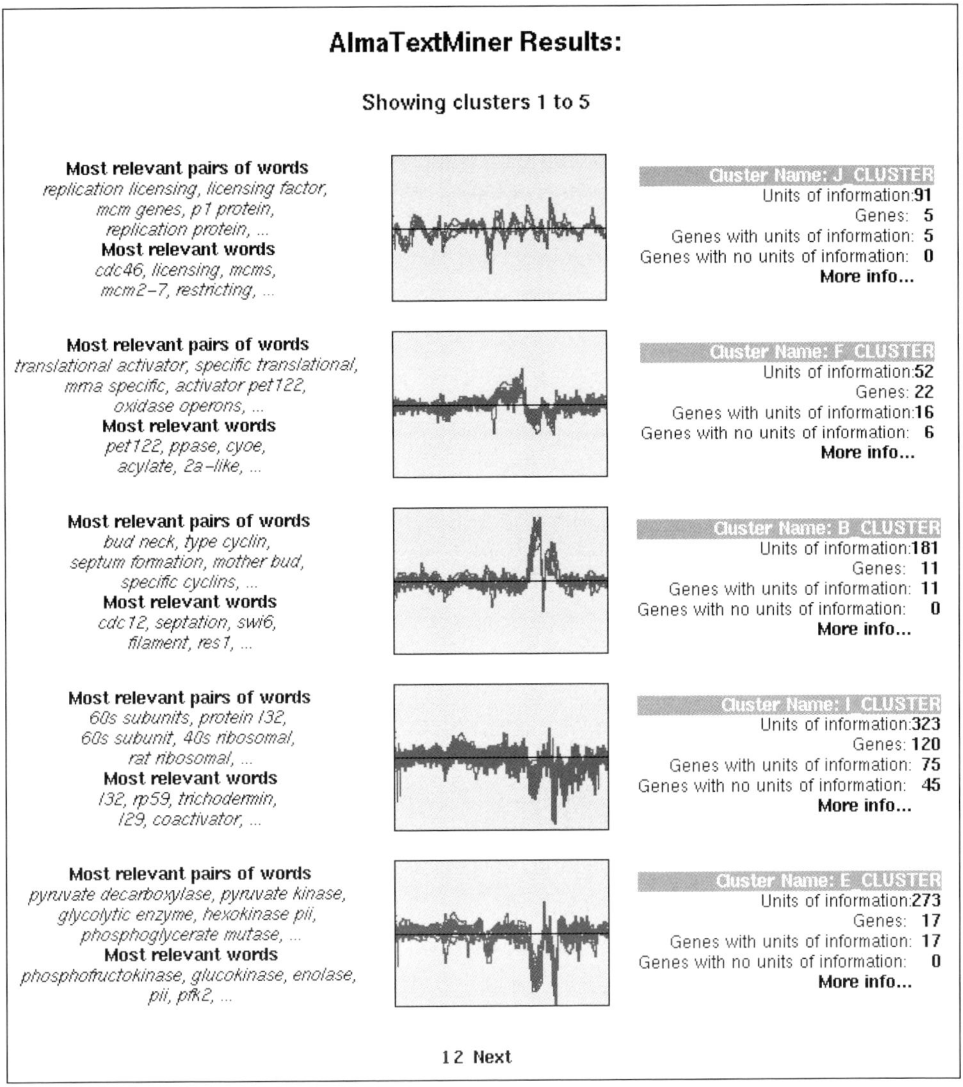

AlmaTextMiner Results:

Showing clusters 1 to 5

Most relevant pairs of words
replication licensing, licensing factor, mcm genes, p1 protein, replication protein, ...
Most relevant words
cdc46, licensing, mcms, mcm2-7, restricting, ...

Cluster Name: J_CLUSTER
Units of information:**91**
Genes: **5**
Genes with units of information: **5**
Genes with no units of information: **0**
More info...

Most relevant pairs of words
translational activator, specific translational, mrna specific, activator pet122, oxidase operons, ...
Most relevant words
pet122, ppase, cyoe, acylate, 2a-like, ...

Cluster Name: F_CLUSTER
Units of information:**52**
Genes: **22**
Genes with units of information:**16**
Genes with no units of information: **6**
More info...

Most relevant pairs of words
bud neck, type cyclin, septum formation, mother bud, specific cyclins, ...
Most relevant words
cdc12, septation, swi6, filament, res1, ...

Cluster Name: B_CLUSTER
Units of information:**181**
Genes: **11**
Genes with units of information: **11**
Genes with no units of information: **0**
More info...

Most relevant pairs of words
60s subunits, protein l32, 60s subunit, 40s ribosomal, rat ribosomal, ...
Most relevant words
l32, rp59, trichodermin, l29, coactivator, ...

Cluster Name: I_CLUSTER
Units of information:**323**
Genes: **120**
Genes with units of information: **75**
Genes with no units of information: **45**
More info...

Most relevant pairs of words
pyruvate decarboxylase, pyruvate kinase, glycolytic enzyme, hexokinase pii, phosphoglycerate mutase, ...
Most relevant words
phosphofructokinase, glucokinase, enolase, pii, pfk2, ...

Cluster Name: E_CLUSTER
Units of information:**273**
Genes: **17**
Genes with units of information: **17**
Genes with no units of information: **0**
More info...

1 2 **Next**

Figure 2. ALMATextMiner is an information extraction system that is able to extract informative keywords from the literature, so helping with the task of annotating function for groups of related genes and assessing the biological relevance of the clusters (http://www.almabioinfo.com/ATM).

Ontologies, descriptions of biological roles and the relationships between them, provide such schema. It is very important to incorporate this information into any tool aimed at mining information from genes and proteins. Mapping the results of the experiment onto ontologies groups together related functions and makes it faster and easier to extract conclusions and to transfer and compare data between different experiments.

The Gene Ontology project (http: //www.geneontology.org) (Ashburner et al, 2000) is probably the best known and most complete ontology available for molecular biology. The ambitious goal of this project is to provide "a tool for the unification of biology", deriving an ontology valid for any organism. Nevertheless, other schema for functional classification have been proposed (see for instance the MIPS classification schema, http: // mips.biochem.mpg.de), that can be very valuable depending on the objectives and preferences of the researcher. We must also take into account that, as ontologies are continously growing and evolving (at least at the present), adding new information and keeping track of the changes is very important. Therefore semi- or fully automatic procedures are needed to perform such task.

Conclusions

Microarray research is already producing a revolution in the study of living systems. Gene expression, organization of celullar processes, responses to different stimuli, etc., can be studied readily. Nevertheless, bioinformatics solutions are needed at every stage of the experiment, and are crucially important for the proper management of the experiments and interpretation of the results. In this chapter we have presented the application of bioinformatics tools to the analysis of expression data at different levels.

Acknowledgements

J.T. wishes to acknowledge Dominic Clark, Juan Carlos Oliveros (Alma bioinformatics), Joaquín Dopazo (CNIO, Madrid, Spain), Christian Blaschke and Alfonso Valencia (CNB, Madrid, Spain) for their help in the preparation of this chapter.

References

Akman, L., and Aksoy, S. 2001. A novel application of gene arrays: *Escherichia coli* array provides insight into the biology of the obligate endosymbiont of tsetse flies. Proc. Natl. Acad. Sci. USA 98: 7546-7551.

Ashburner, M., Ball, C. A., Blake, J. A., Botstein, D., Butler, H., Cherry, J. M., Davis, A. P., Dolinski, K., Dwight, S. S., Eppig, J. T., *et al.* 2000. Gene ontology: tool for the unification of biology. 25: 25-29.

Behr, M. A., Wilson, M. A., Gill, W. P., Salamon, H., Schoolnik, G. K., Rane, S., and Small, P. M. 1999. Comparative genomics of BCG vaccines by whole-genome DNA microarray. Science 284: 1520-1523.

Brown, M. P., Grundy, W. N., Lin, D., Cristianini, N., Sugnet, C. W., Furey, T. S., Ares, M. J., and Haussler, D. 2000. Knowledge-based analysis of microarray gene expression data by using support vector machines. Proc. Natl. Acad. Sci. USA 97: 262-267.

Cohen, B. A., Mitra, R. D., Hughes, J. D., and Church, G. M. 2000. A computational analysis of whole-genome expression data reveals chromosomal domains of gene expression. Nat. Genet. 26: 183-186.

Cummings, C. A., and Relman, D. A. 2000. Using DNA microarrays to study host-microbe interactions. Emerg. Infect. Dis. 6: 513-525.

Debouck, C., and Goodfellow, P. N. 1999. DNA microarrays in drug discovery and development. Nat. Genet. 21: 48-50.

deRisi, J. L., Iyer, V. R., and Brown, P. O. 1997. Exploring the metabolic and genetic control of gene expression on a genomic scale. Science 278: 680-686.

Eisen, M. B., Spellman, P. T., Brown, P. O., and Botstein, D. 1998. Cluster analysis and display of genome-wide expression patterns. Proc. Natl. Acad. Sci. USA 95: 14863-14868.

Evans, W. E., and Relling, M. V. 1999. Pharmacogenomics: translating functional genomics into rational therapeutics. Science 286: 487-491.

Ferea, T. L., and Brown, P. O. 1999. Observing the living genome. Curr. Opin. Genet. Dev. 9: 715-722.

Fritzke, B. 1994. Growing cell structures - a self-organizing network for unsupervised and supervised learning. Neural networks 7: 1141-1160

Geiss, G. K., Bumgarner, R. E., An, M. C., Agy, M. B., van 't Wout, A. B., Hammersmark, E., Carter, V. S., Upchurch, D., Mullins, J. I., and Katze, M. G. 2000. Large-scale monitoring of host cell gene expression during HIV-1 infection using cDNA microarrays. Virology 266: 8-16.

Golub, T. R., Slonim, D. K., Tamayo, P., Huard, C., Gaasenbeek, M., Mesirov, J. P., Coller, H., Loh, M. L., Downing, J. R., Caligiuri, M. A., *et al.* 1999. Molecular classification of cancer: class discovery and class prediction by gene expression monitoring. Science 286: 531-537.

Gray, N. S., Wodicka, L., Thunnissen, A. M., Norman, T. C., Kwon, S., Espinoza, F. H., Morgan, D. O., Barnes, G., LeClerc, S., Meijer, L., *et al.* 1998. Exploiting chemical libraries, structure, and genomics in the search for kinase inhibitors. Science 281: 533-538.

Hayward, R. E., DeRisi, J. L., Alfadhli, S., Kaslow, D. C., Brown, P. O., and Rathod, P. K. 2000. Shotgun DNA microarrays and stage-specific gene expression in *Plasmodium falciparum* malaria. Mol. Microbiol. 35: 6-14.

Herrero, J., Valencia, A., and Dopazo, J. 2001. A hierarchical unsupervised growing neural network for clustering gene expression patterns. Bioinformatics 17: 126-136.

Ideker, T., Thorsson, V., Ranish, J. A., Christmas, R., Buhler, J., Eng, J. K., Bumgarner, R., Goodlett, D. R., Aebersold, R., and Hood, L. 2001. Integrated genomic and proteomic analyses of a systematically perturbed metabolic network. Science 292: 929-934.

Iyer, V. R., Eisen, M. B., Ross, D. T., Schuler, G., Moore, T., Lee, J. C., Trent, J. M., Staudt, L. M., Hudson, J. J., Boguski, M. S., *et al.* 1999. The transcriptional program in the response of human fibroblasts to serum. Science 283: 83-87.

Kohonen, 1990. The self-organizing map. Proc. I.E.E.E. 78: 1464-1480.

Kozian, D. H., and Kirschbaum, B. J. 1999. Comparative gene-expression analysis. Trends Biotechnol. 17: 73-78.

Leemans, R., Loop, T., Egger, B., He, H., Kammermeier, L., Hartmann, B., Certa, U., Reichert, H., and Hirth, F. 2001. Identification of candidate downstream genes for the homeodomain transcription factor Labial in Drosophila through oligonucleotide-array transcript imaging. Genome Biol 2: research0015.0011-0015.0019.

Loftus, S.K., Chen, Y., Gooden, G., Ryan, J.F., Birzniers, G., Hilliard, M., Baxevanis, A.D., Bittner, M., Meltzer, P., Trent, J., and Pavan, W. 1999. Informatic selection of a neural crest-melanocyte cDNA set for microarray analysis. Proc. Natl. Acad. Sci. USA 96: 9277-9280

Marton, M. J., deRisi, J. L., Bennett, H. A., Iyer, V. R., Meyer, M. R., Roberts, C. J., Stoughton, R., Burchard, J., Slade, D., Dai, H., *et al.* 1998. Drug target validation and identification of secondary drug target effects using DNA microarrays. Nat. Med. 4: 1293-1301.

Richmond, C. S., Glasner, J. D., Mau, R., Jin, H., and Blattner, F. R. 1999. Genome-wide expression profiling in *Escherichia coli* K-12. Nucleic Acids Res. 27: 3821-3835.

Rockett, J.C., Luft, J.C., Garges, J.B., Krawetz, S.A., Hughes, M.R., Kirn, K.H., Oudes, A.J., Dix, D.J. 2001. Development of a 950-gene DNA array for examining gene expression patterns in mouse testis. Genome Biol. 2: research0014.1-0014.9.

Scherf, U., Ross, D. T., Waltham, M., Smith, L. H., Lee, J. K., Tanabe, L., Kohn, K. W., Reinhold, W. C., Myers, T. G., Andrews, D. T., *et al.* 2000. A gene expression database for the molecular pharmacology of cancer. Nat. Genet. 24: 236-244.

Sneath, P.H.A. and Sokal, R.R. 1973. Numerical Taxonomy, San Francisco: Freeman.

Tamayo, P., Slonim, D., Mesirov, J., Zhu, Q., Kitareewan, S., Dmitrovsky, E., Lander, E. S., and Golub, T. R. 1999. Interpreting patterns of gene expression with self-organizing maps: methods and application to hematopoietic differentiation. Proc. Natl. Acad. Sci. USA 96: 2907-2912.

Tavazoie, S., Hughes, J. D., Campbell, M. J., Cho, R. J., and Church, G. M. 1999. Systematic determination of genetic network architecture. Nat. Genet. 22: 281-285.

Törönen, P., Kolehmainen, M., Wong, G., and Castrén, E. 1999. Analysis of gene expression data using self-organizing maps. FEBS letters 451: 142-146.

Wilson, M., DeRisi, J., Kristensen, H. H., Imboden, P., Rane, S., Brown, P. O., and Schoolnik, G. K. 1999. Exploring drug-induced alterations in gene expression in *Mycobacterium tuberculosis* by microarray hybridization. Proc. Natl. Acad. Sci. USA 96: 12833-12838.

Zhou, Y. X., Kalocsai, P., Chen, J. Y., and Shams, S. 2000. Information processing issues and solutions associated with microarray technology. In: Microarray Biochip Technology. BioTechniques Books. M. Schena, ed. Natick, MA. p. 167-200.

From: *Bioinformatics and Genomes: Current Perspectives*
Edited by: Miguel A. Andrade

Chapter 4

High Rate of Gene Displacement in Vitamin Biosynthesis Pathways

Enrique Morett, Gloria Saab-Rincon, Enrique Merino, Peer Bork, Emmanuvel Rajan, Leticia Olvera, and Maricela Olvera

Abstract

One of the most challenging tasks in genomic science is the prediction of the function of genes for which there is no clear sequence similarity to annotated genes. However, it is even more challenging to assign the correct function to genes that display sequence similarity to genes of unrelated function: analogous enzymes perform the same biochemical reaction but they are not phylogenetically related, such that it is not possible to identify them by sequence similarity. Here we propose that the vitamin biosynthetic pathways have experienced multiple events of gene loss and recovery of function by unrelated genes, the so-called analogous gene displacement. We carried out an extensive search for the genes that participate in thiamin biosynthetic pathways in the completely sequenced genomes. We show that the great majority of these organisms lack from a few to many orthologs to the *Escherichia coli thi* genes. We searched for the analogous enzymes using gene neighbourhood, co-ocurrence in operons, identification of regulatory sequences, and anticorrelation strategies. Our strategy resulted in the identification of some possible analogous enzymes in this pathway.

Introduction

Generally, enzymes that catalyze the same biochemical reaction show a certain degree of sequence similarity and/or are structurally closely related. This implies that, normally, they have a common phylogenetic origin and have evolved by divergence from an ancestral protein. Fitch (1970) coined the term orthologs to name such proteins when they are from different organisms, while paralog proteins denotes the same evolutionary origin but acquisition of new function by duplication and divergence. Paralog genes normally code for enzymes that have related activities and belong to the same structural family. However, since the early ages of enzymology, it has been documented that there are also apparently unrelated enzymes that present the same activity: the so-called analogous enzymes (Warburg and Christian, 1943). But, how common are analogous enzymes in nature?

In a recent analysis of all the enzyme sequences deposited in the public sequence data banks, Koonin and coworkers reported that analogous enzymes are more common that previously thought. They showed that more than 30 reactions are catalyzed by two or more proteins for which neither amino acid sequence nor structural similarity can be detected (Galperin *et al.*, 1998). Thus, it is inferred that these enzymes have then very likely evolved independently rather than from a common ancestral protein, and converged into the same catalytic function.

Is Orthologous Gene Displacement Frequent In Vitamin Biosynthesis Pathways?

The aim of this chapter is to propose that the presence of analogous enzymes is very common in pathways for the biosynthesis of compounds that are required in minute amounts in the cells, as vitamins. Our rationale is that since vitamins are required in so low amounts in the cell, but on the other hand they are absolutely required for cell growth, mutations that impair the activity of enzymes in their biosynthetic pathways could be suppressed by mutations in unrelated enzymes that alter their activity, such that they result in a weak catalytic activity towards the substrate of the affected enzyme. An extremely low catalytic activity of an enzyme that is expressed at high levels could be sufficient to provide the required limited amounts of the vitamin in question. Thus, very inefficient enzymes could provide enough activity and therefore be readily selected with the new activity. Later, via an evolutionary optimization process, the enzyme with the new activity would attain the required levels of activation and expression. It is likely that similar processes could also operate in other pathways, but the low amount of vitamins needed made them more plausible to evolve independently from unrelated enzymes. For example, pathways for secondary metabolism could also be targets of gene displacement. In such pathways the selective pressure to recover the activity of a mutant cell that is unable to synthesize a metabolite, not required for survival, would not be so intense as compared to vitamin biosynthesis.

As an example of the presence of multiple analog enzymes in vitamin biosynthetic pathways we present our analysis of the thiamin biosynthesis in the completely sequenced microbial genomes. It is shown that the great majority of organisms lack from one to several genes orthologs to the *thi* genes of *E. coli*. As fast divergence of enzymes in vitamin biosynthesis is rather unlikely, the absence of identifiable orthologs indicates that multiple events of gene displacement, that is the presence of analogous enzymes, have occurred in the biosynthetic pathway of thiamin.

Thiamin, also known as vitamin B1, is a cofactor for several key enzymes involved in carbohydrate metabolism, therefore it is essential for growth (White and Spencer, 1996). Thiamin phosphate is made by the condensation of hydroxy methylpyrimidine pyrophosphate (HMP-PP) and methyl (-hydroxyethyl) thiazole phosphate (THZ-P) moieties, by the enzyme thiamin phosphate synthase, coded by the *thiE* gene (see Figure 1). Thiamin phosphate is converted to the pyrophosphate form, the active form of the vitamin, by thiamin kinase, the product of the *thiL* gene. In *E. coli* at least two gene products, *thiC* and *thiD* participate in the formation of HMP-PP from AIR, an intermediate in purine biosynthesis, while at least six gene products, *thiH*, *thiI*, *thiS*, *thiG*, *thiF* and *thiM*, are needed for the synthesis of THZ-P (Begley *et al.*, 1998).

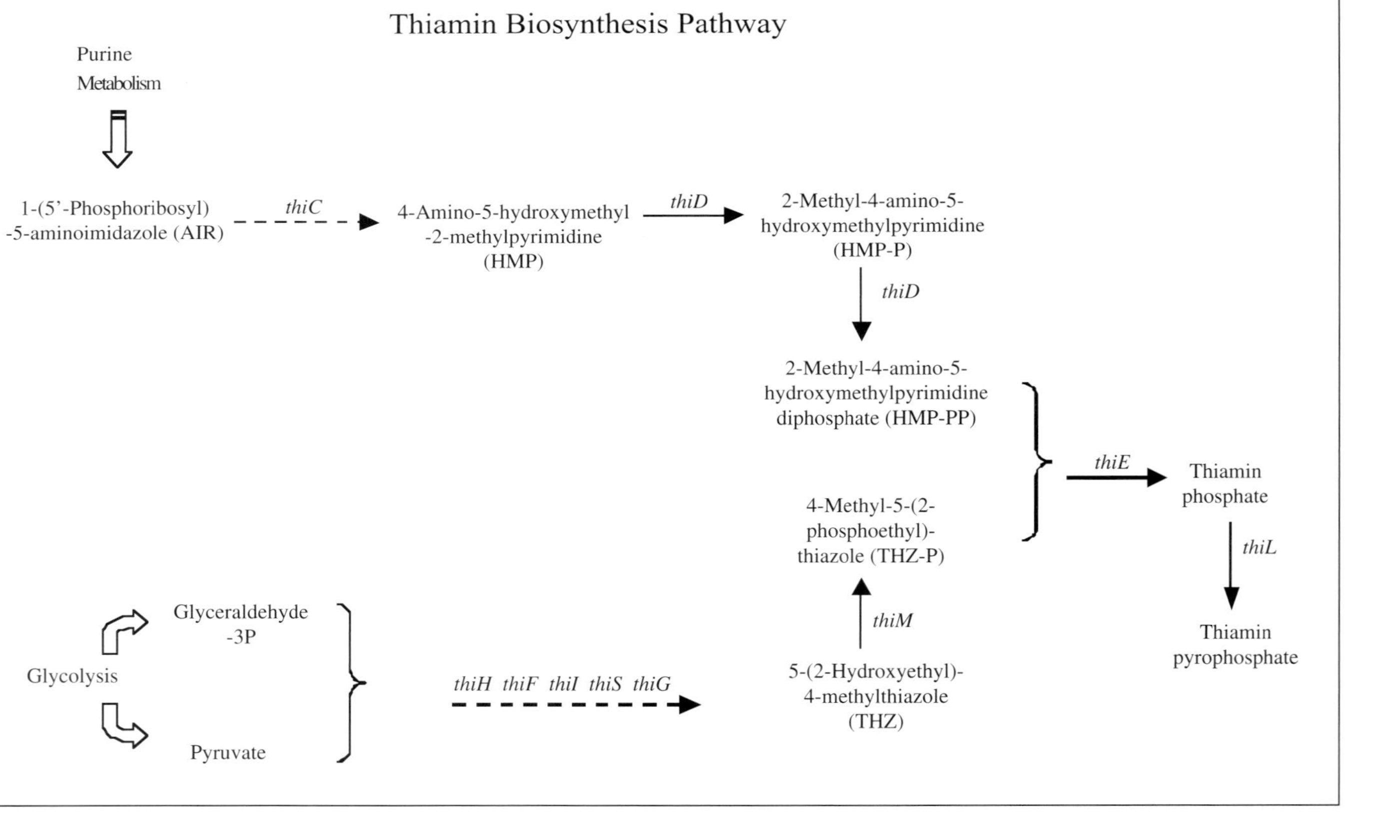

Figure 1. *E. coli* thiamin biosynthesis pathway. Discontinuous arrows indicate unknown parts of the pathway. The functional implication of some genes to those unknown steps might be known (e.g. *thiC* is known to be somehow involved in the transformation of AIR into HMP, and it is not kwnown whether there are other proteins involved in the process). Thick arrows indicate connections with other important pathways. Note that the product of *thiD* catalyses two different consecutive steps.

Search For Orthologs Of The *E. coli thi* Genes In The Completely Sequenced Genomes By Sequence Similarity

We searched for the presence of ortholog genes to the *E. coli thi* genes in the complete microbial genomes, and analysed their genomic organization. Nine Archeal, 33 Eubacterial and the genomes of the yeast *Saccharomyces cerevisiae* and *Candida albicans* were analysed. It follows a list of the genes found in these completely sequenced genomes.

thiD is the gene that is present in more of the organisms analysed; it is present in all microbes but *Synechocystis* including those with small genomes, *Rickettsia*, *Chlamydia*, *Ureaplasma*, and *Mycoplasma,* which lack almost all the enzymes for *de novo* thiamin synthesis. The ThiD enzyme catalyzes the two consecutive phophorylation steps in the synthesis of HMP-PP. The products of *thiJ* and *pdxP* can also phosphorylate HMP to the mono phosphate form, but these proteins are involved mainly in scavenge of HMP from the media and we will not discuss them further.

thiE and *thiL* are the second most ubiquitous genes after *thiD*. *thiE* codes for the key enzyme in thiamin biosynthesis, thiamin phosphate synthase. This enzyme condensates the THZ-P and HMP-PP to produce thiamin monophosphate, which is converted to the pyrophosphate form by thiamin kinase, the product of *thiL*. *thiE* is present in yeast and in all the eubacterial species but *Thermotoga maritima,* and is missing in the archeobacteria *Halobacterium, Methanobacterium, Methanococcus, Pyrococcus* and *Aeropyrum.* Like *thiE*, *thiL* is missing in *Thermotoga maritima;* it is as well missing in *Deinococcus radiodurans*, *Lactococcus*, and *Streptococcus*. It is present in all the Archea but it is missing in yeast and *Candida*. Sequence similarity using PSI-BLAST searches (Altschul *et al.*, 1997) failed to detect orthologs of *thiL* in other Eukaryotes. It is very likely a gene only present in Archea and Eubacteria.

thiC is present in the three domains of life, being missing only in *Thermoplasma*, *Aeropyrum pernix*, *Lactococcus lactis*, *Streptococcus pyogenes*, *Haemophilus influenzae*, and *Pasteurella multocida*. The function of its product is not yet known. *thiM* is scattered in the three domains of life. It is present in the Archea *Archaeoglobus fulgidus* and *Pyrococcus*, and in yeast. In Eubacteria it is present in the firmicutes, *Enterobacteriae*, *Pasteurellacea*, *Helicobacter*, and *Campylobacter*. ThiM phosphorylates THZ to THZ-P.

thiS codes for a very short protein of 66 amino acids which has some sequence similarity with MoaD. Both proteins have two glycines at the COOH terminus that are involved in sulfur transfer via the formation of a thiocarboxylate at the COOH end of the protein. ThiS donates the sulfur for the formation of thiazole in a reaction catalyzed by both ThiI and IscS proteins. *thiS* is present in the Archea *Archaeoglobus* and *Methanobacterium*. It is widely present in Eubacteria although it is missing in *Thermotoga, Aquifex,* and *Haemophilus.*

thiF is highly similar to *moeB*, a gene involved in the biosynthesis of the molibdopterin cofactor. The high level of similarity has lead to several miss-annotations in many organisms. We identified unambiguously *thiF* genes not only by their higher sequence similarity to *E. coli thiF* than to *E. coli moeB*, but also by being located in operons with other *thi* genes in the following organisms: *Campylobacter, Bacillus, Escherichia coli, Aquifex, Neisseria, Vibrio, Xylella, Helicobacter, Archaeoglobus, Aeropyrum*, and *Methanobacterium.* ThiF is involved also in the sulfur transfer process to form THZ. It has recently been found that a covalently linked protein-protein conjugate is formed between ThiF and ThiS similar to the ubiquitin-E1 conjugate (Xi *et al.*, 2001).

thiI has a dual role in *Escherichia coli.* Its product is not only involved in thiamin biosynthesis, transferring sulfur to ThiS, but also participates in the formation of thiouridine in response to UV light, a protective mechanism that allows the cells to survive to the DNA damage. *thiI* is found in many organisms including the small genomes but *Chlamydia.* However, we did not find *thiI* homologues in *Campylobacter, Aquifex, Mycobacterium, Neisseria, Xylella, Helicobacter,* and *Haemophilus,* although almost all these organisms have *thiS,* as shown above. Then, there must be a different protein involved in the sufphur transfer to ThiS in these bacteria.

thiG is only present in Eubacteria but no orthologs were found in *Thermotoga, Lactococcus,* and *Haemophilus.* The exact role of ThiG is still not clear. *thiH* is the least represented gene. However, Eubacteria as diverse as *Thermotoga, Escherichia, Vibrio*, and *Helicobacter* have this gene. *thiH* has sequence similarity to several genes involved in the biosynthesis of sulfur containing cofactors such as biotin (*bioB*), lipoic acid (*lipA*), PQQ (*pqqE*), and molibdopterin (*moaA*). The products of all these genes share a highly conserved cysteine-rich motif. These observations suggest that a similar sulfur incorporation takes place in the synthesis of all these vitamins.

Search For *thi* Genes Not Present In *E. coli*

As shown above, many organisms prototrophs for thiamin lack from one to several of the known *thi* genes. Actually, only the organisms closely related to *E. coli* show the complete set. This could imply that they utilize alternative pathways for thiamin synthesis or that many events of non-orthologous gene displacements have occurred and the *thi* genes have been reinvented several times in nature.

How the nonortholog genes could be identified? Methods of finding these missing genes include the search for physical linkage of open reading frames (Lathe *et al.*, 2000) and patterns of ortholog co-ocurrence in different species (Pellegrini *et al.*, 1999). Also, the presence of highly conserved regulatory sequences, the *thi* box (Miranda-Rios *et al.*, 1997; 2001), could be used to detect specifically *thi* genes. Using these tools, we have identified several ORFs that might be involved in thiamin biosynthesis.

Search For *thi* Genes In Complete Genomes Using Gene Neighbourhood And Anticorrelation Of Occurrence

In the search for analog genes of *thiE* we found a gene (that we denote *thiE**) present in the Archeal as well as *Thermotoga*, organisms in which we could not find a *thiE* homologue gene. We did not find any homolog of *thiE** in organisms that have bona fide *thiE*, except *Pyrococcus.* This gene is actually fused to *thiD* forming a predicted polypeptide of about 400 amino acids. The extended region of this unusual *thiD* gene did not show sequence similarity to any other gene in the public data banks searched using PSI-BLAST. This anticorrelation and the fact that in these organisms, *thiE* is the only gene missing in common in the biosynthetic pathway of thiamin, makes very tempting to speculate that the *thiE** gene codes for an analog thiamin synthase enzyme. It will be very interesting to determine whether this gene can complement a *thiE* mutant strain of *E. coli*. We are currently constructing such a strain and cloning the extended *thiD* gene of *T. maritima* to check this hypothesis.

In *Rhizobium etli* there is a gene, *thiO*, for which there is no ortholog in *E. coli,* that has been proposed to participate in thiamin biosynthesis (Miranda-Rios *et al.*, 1997). This gene forms part of a *thiCOSGE* operon. We searched for ortholog genes to *thiO* in the completely sequenced genomes. We identified likely ortholog genes to *thiO* in *Aquifex, Mycobacterium, Bacillus,* and *Neisseria*. None of these species have *thiH*. Interestingly, *Bacillus subtilis* has almost the complete set of *thi* genes identified in *E. coli*, with the exception of *thiH.* Thus, we found that *thiO* and *thiH* are mutually exclusive; in other words, their presence anticorrelates in the completely sequenced microorganisms. Thus, we speculate that *thiO* and *thiH* have equivalent functions being a case of non-ortologous gene displacement. Neither the function of the *thiO* product nor the function of *thiH* gene product is known. Both *thiO* and *thiH* are distributed among different Eubacterial taxa. Then it is difficult to predict which gene is the ancestral *thi* gene. Interestingly neither Archea nor the microbial Eukaryotes have any of these genes, implying that there must be a third analog gene coding for the same function in these organisms. We are constructing an *E. coli* strain devoid of *thiH* to try to complement its function with *thiO* from *R. etli*.

Search For The Regulatory *thi* Box

The *thi* box is a conserved regulatory sequence of about 38 nucleotides present in front of several *thi* genes in both Eubacteria and Archea organisms. About one half of the positions of the *thi* box are almost invariant, but all can fold in a similar RNA structure due to correlated substitutions that maintain base-pair complementarity (Miranda-Rios *et al.*, 2001). It has been proposed that this sequence, once transcribed to mRNA, can directly bind thiamin and down-regulate the expression of *thi* genes when the vitamin is present. No other regulatory sequence is conserved in such diverse group of bacteria.

We searched for possible *thi* boxes in the leader region of all the genes of the completely sequenced genomes in an attempt to identify new putative analog *thi* genes. We developed a program that scans the 500 nt sequence upstream of every gene, according to the GenBank database annotations. A 30 nt sliding window is used to search for the acctga conserved sequence in the 5' side of the *thi* box element. If a window has a perfect match or only one mismatch, then the program searches for the cgnngg sequence at the 3' end of the *thi* box element. Since the analysis of *thi* box sequences present in different genomes revealed that the 3' end sequence may vary importantly, our program allows the following variants: gnngg, gnnngg, cnnngg, cnnnngg, cgnnng, cgnnnng, cgnng, cgnnng. If both 3' and 5' conserved strings are found, then the program predicts the secondary structure of the 30 nt window using the FoldRNA program of the GCG package. In case that the predicted mRNA fold corresponds to a stem-and-loop structure the program verifies that its free energy is smaller than zero. If the structure meets this requirement, the program verifies that all of the bases found in the 5' (acctga) and 3' (cgnngg) searches, are part of a stem or a loop, accordingly with its position within the consensus *thi* box mRNA fold. If a window meets all of these considerations, its sequence is therefore considered as a putative *thi* box element.

Our search strategy resulted in the identification of several of the *thi* boxes used to construct the searching matrix and of *thi* boxes corresponding to already identified *thi* genes for which no *thi* box had previously been described. Additionally, several other genes having *thi* boxes were found.

Several seem to be false positives, since they are involved in non vitamin-related functions. In general, these *thi* boxes showed a low score. However, many other quite interesting hits were obtained which support a common role of thiamin in the regulation of at least some genes involved in the molibdopterin cofactor and biotin biosynthesis. Then, we speculate that thiamin can, in some organisms regulate the expression not only of the genes involved in its synthesis but also of genes involved in other sulfur-containing vitamins. Following we describe some of the most relevant examples.

The *ydbH* gene of *Lactococcus* presents at its untranslated region a highly probable *thi* box. This gene codes for a protein of unknown function but it is located in an operon with *bioY* which codes for biotin synthase. The *Thermoplasma acidophilum* gene *Ta0442* also has a putative *thi* box. This gene is in an operon with the molibdopetrin synthesis gene *moaA*.

Another gene for which we identifyed a *thi* box is the *ylmB* gene of *B. subtilis.* This gene is paralog to *argE,* and in *B. halodurans* the ortholog of this gene is in an operon with *thi4*, a gene first identified in *Neurospora crassa* as involved in thiamin biosynthesis. It is very intriguing that *B. subtilis ylmB* has a *thi* box and that its ortholog in the closely related *B. halodurans* species is in an operon with a thiamin related gene, although we did not detected a *thi* box in the latter gene.

Interestingly, we found *thi* boxes in the intergenic region of mRNAs. For example there is a *thi* box exactly before the start point of the *B. subtilis thiO* gene. This gene is the third gene in an operon with other *thi* genes. In *Synechocystis* the *brkB* gene is the second gene in an operon with *thiC* and there is a thi box in front of it. The *Thermoplasma* species have a gene with sequence similarity to membrane transporters, TVN1053, which is the second gene in an operon with *moaA*. The fact that we found *thi* boxes in these intergenic regions strongly supports the hypothesis that these regulatory sequences exert their role at the level of translation (Miranda-Rios *et al.*, 2001). The lack of terminator sequences and the low stability of the *thi* boxes makes them unlikely to play a role in transcriptional termination.

Conclusions

In this chapter we analysed the thiamin biosynthesis pathways in the completely sequenced microorganisms. Almost all of them lack at least one of the *thi* genes reported for *E. coli*. The fact that the majority of these organisms do not require thiamin for growth indicates that there are multiple analogous genes for the biosynthesis of this vitamin. Some genes likely to participate in this process were identified by gene neighbourhood, co-occurrence, and for the presence of *thi* boxes. These strategies can certainly be used not only for the identification of genes that participate in vitamin biosynthesis but also for the identification of many other cellular functions in the completely sequenced organisms.

Acknowledgments

This work has been supported in part by grants from CONACyT and DGAPA, UNAM. EM was an Alexander von Humboldt Fellow.

References

Altschul, S.F., Madden, T.L., Schaffer, A.A., Zhang, J., Zhang, Z., Miller, W., and Lipman, D. 1997. Gapped BLAST and PSI-BLAST: a new generation of protein database search programs. Nucleic Acids Res. 25: 2289-3402.

Begley, T.P., Downs, D.M., Ealick, S.E., McLafferty, F.W., Van Loon, A.P., Taylor, S., Campobasso, N., Chiu, H.J., Kinsland, C., Reddick, J.J., and Xi, J. 1999. Thiamin biosynthesis in prokaryotes. Arch. Microbiol. 171: 293-300.

Galperin, M.Y., Walker, D.R., and Koonin, E.V. 1998. Analogous enzymes: independent inventions in enzyme evolution. Genome Res. 8: 779-790.

Fitch, W.M. 1970. Distinguishing homologous from analogous proteins. Syst. Zool. 19: 99-113.

Lathe, W.C. 3rd, Snel, B., and Bork, P. 2000. Gene context conservation of a higher order than operons. Trends Biochem. Sci. 25: 474-479.

Miranda-Rios, J., Morera, C., Taboada, H., Davalos, A., Encarnacion, S., Mora, J., and Soberon, M. 1997. Expression of thiamin biosynthetic genes (thiCOGE) and production of symbiotic terminal oxidase cbb(3), in *Rhizobium etli.* J. Bacteriol. 179: 6887-6893.

Miranda-Rios, J., Navarro, M., and Soberon M. 2001. A conserved RNA structure (thi box) is involved in regulation of thiamin biosynthetic gene expression in bacteria. Proc. Natl. Acad. Sci. USA. 98: 9736-9741.

Pellegrini, M., Marcotte, E.M., Thompson, M.J., Eisenberg, D., and Yeates, T.O. 1999. Assigning protein functions by comparative genome analysis: protein phylogenetic profiles. Proc. Natl. Acad. Sci. USA. 96: 4285-4288.

Warburg, O., and Christian, W. 1943. Isolierung und kristallization des garungsferments zymohexase. *Biochem. Z.* 314: 149-176.

White, R.L., and Spencer, I.D. 1996. Biosynthesis of thiamin. In: *Escherichia coli* and *Salmonella*. Cellular and Molecular Biology. F.C. Neidhard *et al.*, eds. ASM Press, Washington. p. 680-686.

Xi, J., Ge, Y., Kinsland, C., McLafferty, F.W., and Begley, T.P. 2001. Biosynthesis of the thiazole moiety of thiamin in *Escherichia coli*: identification of an acyldisulfide-linked protein-protein conjugate that is functionally analogous to the ubiquitin/E1 complex. Proc. Natl. Acad. Sci. USA. 98: 8513-8518.

From: *Bioinformatics and Genomes: Current Perspectives*
Edited by: Miguel A. Andrade

Chapter 5

Prediction of Post-Translational Modifications From Amino Acid Sequence: Problems, Pitfalls, and Methodological Hints

Frank Eisenhaber, Birgit Eisenhaber and Sebastian Maurer-Stroh

Abstract

Although high-throughput techniques for experimental proteome characterization are still in their infancy, the number of reported instances of post-translational modifications of proteins is already larger than the number of their sequences in databases. Thus, a typical protein appears to be covalently modified several times during its lifetime, to allow the regulation of its function. Given the huge number of sequences of otherwise uncharacterised proteins resulting from genome projects, the computer-aided prediction of the possibility of post-translational modifications from amino acid sequence becomes a necessity for genome annotation.

Only some types of post-translational modification can be considered as reasonable targets for predictor development for the moment due to problems with the quality of learning sets and the complexity of protein substrate recognition for modification. Information on the sequence motif responsible for targeting can be extracted from the sequence variability of natural substrate proteins as well as of model compounds and from the structures of responsible enzymes. The motif description needs to be rich enough, albeit not necessarily in terms of positional amino acid type preferences, for a reliable discrimination of true substrates from unrelated sequences. Solutions for the construction of prediction functions as well as for the assessment of false prediction with rigorous statistical criteria are reviewed.

Protein Function And Post-translational Modifications Of Proteins

Whereas the genome represents only the entirety of potential cellular development programs, information on the proteome as its executed portion at a given time is the really relevant desired data in scientific applications. The nucleic acids mainly fulfil the tasks of storage and transfer of genetic information in living organisms. At the same time, the proteins form complicated cellular machineries for realization of this genetic program in dependency and in response on changing environment conditions. Since the experimental determination of complete genome sequences has become a solved technical problem and, as a consequence, the fraction of functionally non-characterized genes in databases continues to grow, determination of protein function both with experimental means and theoretical approaches is the next, most important task.

Post-translational modifications (PTMs) have a great impact on protein size, hydrophobicity, and other physico-chemical properties. Therefore they can change, enhance, or block specific protein activities, or target the protein to another subcellular localization. Thus, PTMs directly influence protein function. Regrettably, the term "protein function" is only weakly characterized at the textbook level. Its definition has to reflect the complexity of living systems and requires a hierarchical description (Bork *et al.*, 1998; Eisenhaber and Bork, 1998):

i) The presence and activity of a gene product may be directly associated with a known *phenotypic function* (or *phenotypic disfunction* in the case of gene loss) but, in most cases, phenotypic properties are determined by the co-operative action of many gene products and the influence of a specific protein is hardly characterized in quantitative terms.

ii) A set of many co-operating proteins is responsible for a *cellular (physiological) function* (metabolic pathway, signal transduction cascade, cytoskeletal complex, etc.). The cellular function of a protein is always context-dependent and is characterized by taxon, organ, tissue, etc. Cellular function has a number of attributes: the amount of protein molecules is controlled via gene expression which might be limited to certain types of cells or tissues or to specific periods in the cell cycle or the individual ontogenese (expression pattern). Also, translational regulation influences protein copy number. In contrast to early views, proteins are not static entities after completion of mRNA translation. Typically, the protein activity is tightly regulated as a result of interaction with other biomacromolecules (ligand binding), via processing (covalent PTM), and sorting (subcellular localization: translocation to the correct intra- or extracellular compartments).

iii) Protein function at the molecular level is rather a list of potential capabilities determined by its primary and tertiary structures. Molecular function description includes qualitative and quantitative aspects of diffusion properties, conformational flexibility, allosteric conformational changes, possible ligand-binding (or catalytic) activities and ability for PTMs (polypeptide excision, protein splicing or modification of single residues). Depending on cellular context, different features of the molecular function may become important.

With PTMs, proteins can go beyond the limitations in chemical structure imposed by the set of 20 natural amino acid monomers and, therefore, may assume a much larger variety of complementary functional properties. Prediction of PTMs from amino acid sequence is an important academic task that is critical for the biological interpretation of genome data as has been demonstrated in the case of the signal leader peptides (Nielsen *et al.*, 1999), GPI-lipid anchor sites (Eisenhaber *et al.*, 2000), and protein myristoylation (Maurer-Stroh *et al.*, 2002b). It should be noted that the apparatus for the prediction of PTMs is sometimes also applicable in the case of sorting signals such as the peroxisomal targeting motif (Neuberger *et al.*, to be published).

In this chapter, we discuss general aspects of method development in PTM prediction. (a) Since recognition of potential PTMs in the protein sequences *ab initio* does not appear promising, todays attempts for PTM prediction have to rely on knowledge-based approaches. Therefore, issues of learning set collection and evaluation are of principal importance. Obviously, SWISS-PROT (Bairoch and Apweiler, 2000) is the prime source for annotated protein sequences. A survey of PTMs described in SWISS-

PROT was the starting point of the work described in this chapter. To our surprise, we found more instances of reported PTMs than protein sequences. (b) Then, we determine conditions for PTMs that might be targets for successful predictor development. We will see that the scarcity of available sequence data and heterogeneity of additional experimental information puts specific requirements on the predictor construction. (c) A high rate of correct recognition of known exemplary cases of the PTM considered is by far not sufficient for a prediction tool in genome annotation applications. The false positive prediction rate calculated over large relevant datasets should be low enough (<<5%). Rigorous statistical criteria for different sequential localizations of the predicted PTMs are exemplified in the last section. Here, we generalize our experience accumulated during the work on GPI-lipid anchor prediction (Eisenhaber *et al.*, 1999; 2001) and myristoyl anchor prediction (Maurer-Stroh *et al.*, 2002a; 2002b).

Definition And Classification Of Post-Translational Modifications Of A Protein

The review of Uy and Wold (1977) is an early peak of public awareness with respect to post-translational covalent modifications of proteins. They listed 140 modified versions of the 20 common amino acids. However, this list also included "pre-translationally" modified amino acids (while or even before they are bound to tRNA) as well as artefacts resulting from harsh experimental processing of proteins. Lists of the various molecular fragments from proteins undergoing extensive post-translational processing can also be retrieved from modern mass spectrometer manuals (> 300 different amino acid modifications).

However in this chapter, we will deal with the naturally occurring PTMs that have an impact on the physico-chemical properties of proteins and, therefore, modify specific protein functions *in vivo*, for example, targeting the protein to another subcellular localization. In addition, we should emphasize at this point that naturally occuring PTMs are certainly not limited to a chemical change of one single amino acid residue.

An exact classification of the extensively growing number of known PTMs is difficult, also due to the lack of a clear definition. Simple derivation from the name itself results in too loose a definition: not any change of a protein is a PTM in the strict sense. Philosophically, it could be argued that even the folding of the nascent protein into its three-dimensional structure is a sort of post-translational modification. Accordingly, since proteins are highly dynamic structures, every later conformational change could be considered as such. However, the driving forces along the protein folding and unfolding pathways result mainly from non-covalent interactions. Therefore, a necessary restriction for PTMs is the covalent modification of the protein.

On the other hand, not every new formation of a covalent bond should be seen as a PTM since this definition alone would also include an enormous number of interactions of specialized proteins with their very specific binding partners. A generality requirement is, therefore, reasonable. The considered type of covalent modification has to be a general feature of proteins, not only specific for a minor subclass.

It should be emphasized that not only the formation of covalent bonds but also their cleavage complies with this definition. Different types of proteolytic processing of proproteins, including protein splicing, can be seen as a PTM since they have a deterministic influence on size, properties, localization and function of very different proteins. Hence, PTMs can be classified into groups from the chemical point of view:

1) Cleavage of covalent bonds including the breakage of a peptide bond and the removal of groups from single amino acid residues.
2) Formation of covalent cross-links of intra- or intermolecular (with a ligand/another protein) nature. The covalent attachment of cofactors (prosthetic groups) is a subset of the latter case.
3) Combinations of (1) and (2).

Furthermore, PTMs can occur as a result of either an enzymatic process or a spontaneous reaction. In the first case, binding of the substrate protein at the binding site of the enzyme over a number of sequence positions near the modification site is a critical factor in the recognition motif specificity. In this chapter, we consider only the prediction of such PTMs. Spontaneous PTMs occur under the sole involvement of the reactants themselves. Except for cases of autocatalysis (for example, asparagine deamidation (Robinson and Robinson, 2001)), the second agents are often metabolites (glucose, urea, etc.) or xenobiotics (Harding, 1985). Many of these reactions are detrimental for the cells, which are apparently optimised to suppress them. Since the reaction rate of such spontaneous PTMs is normally low, they occur to long-living proteins, or through pathological processes where the reactants reach abnormally high concentrations (Harding, 1985; Lee and Cerami, 1992; Rattan *et al.*, 1992).

ANNOTATION OF POST-TRANSLATIONAL MODIFICATIONS IN SWISS-PROT

Among all of the available protein sequence databases, SWISS-PROT is the currently best annotated one regarding a variety of post-translational modifications. There are only a few, very recently discovered cases of PTMs, such as protein S-nitrosylation (Jaffrey *et al.*, 2001), for which the scientific literature alone appears a dramatically richer source of information. We extracted information from the SWISS-PROT feature table, the keyword, and the comment lines (with tokens FT, KW, and CC, respectively). The detailed results derived from SWISS-PROT release 40 are listed in Table 1. As the descriptions of the particular PTMs applied by the authors of the entries are not consistent throughout the database, regular expressions for each individual PTM have been refined to detect as many protein entries mentioning a PTM as possible. Therefore, the numbers in Table 1 can be considered as quite accurate estimates of occurrences of annotated PTMs in SWISS-PROT.

Obviously, it is unlikely that the relative occurrences of the PTMs in Table 1 reflect their frequency in nature. On the one hand, the annotation instances for a given PTM are often not directly comparable, since the criteria that have been applied by the authors of the sequence entries in SWISS-PROT are not uniform. (It is also known that several real annotation errors are contained in the database but this is also not our major reason for concern). On the other hand, we must assume that the relative distribution of occurrences is biased by the ease of experimental determination of the specific PTM, the current knowledge of its biological importance, the existence of characterized homologues, or the knowledge of apparently simple amino acid type patterns for the recognition motif. Therefore, what looks like a curiosity due to the small number of instances nowadays, might turn out to be a much more common modification than previously thought when further experimental data is available. Moreover, it is clear that not all naturally occurring PTMs have yet been discovered.

Although only modifications occurring more than 10 times were included in Table 1, the list still contains 76 different entries. Strikingly, the overall number of annotated post-translational modifications (119401) already exceeds the total number of sequences in SWISS-PROT (101602)! As soon as complete genomes are available and the biological interpretation of the sequences becomes the major task, the often underestimated post-translational protein processing will receive much more attention, simply due to the fact that many proteins appear to be subjected to several modifications during various stages of their life time.

Table 1. Occurrences of posttranslational modifications in sequence annotation of SWISS-PROT release 40. The occurrences of post-translational modifications as annotated in the FT-lines, and if not in the FT-, then in the CC- and KW-lines have been counted. We list the total number and the values separately by annotation quality comments (potential or probable, by similarity, or other comments). Only modifications reported for 10 or more different, independent protein instances are reported.

COVALENT BOND	POT./PROB.	BY SIM.	OTHER	TOTAL
CLEAVAGE				
SIGNAL PEPTIDES	7457	1446	4056	12959
INITIATING METHIONINE	53	1890	2949	4892
PROPEPTIDE	1002	943	1897	3842
TRANSIT PEPTIDE	1016	499	785	2300
PEPTIDE	134	130	1938	2202
CLEAVAGE SITE	110	184	240	534
PROTEIN SELF SPLICING (INTEIN)	38	5	8	51
BREAKPOINT FOR TRANSLOCATION FORMATION	0	0	35	35
INTRAMOLECULAR				
DISULFIDE BRIDGES	1455	27458	4427	33340
THIOETHER	0	19	71	90
LANTHIONINE	0	3	60	63
THIOLESTER	0	9	14	23
CYCLIZATION				16
INTERMOLECULAR				
CARBOHYDRATES				40298
N-LINKED	37192	324	1487	39003
GLCNAC	37046	324	1458	38828
HIGH MANNOSE	5	0	22	27
GALNAC	18	0	2	20
O-LINKED	256	250	745	1251
GAL	29	83	451	563
GLYCOSAMINOGLYCAN	141	24	12	177
GLC	8	79	48	135
MAN	1	4	47	52
FUC, HEXNAC or XYL	0	6	12	28
C-LINKED (MAN)	1	5	36	42
PHOSPHORYLATION	1162	2791	1001	4963
LIPID				2679
PALMITATE	287	524	92	903
N-ACYL DIGLYCERIDE	440	36	54	530
GERANYL-GERANYL	26	424	41	491
MYRISTATE	60	240	101	401
GPI-ANCHOR	108	73	41	222
FARNESYL	15	70	47	132
ACETYLATION	108	629	788	1525
INTERCHAIN DISULFIDE BONDS	341	713	462	1516
AMIDATION	111	116	1138	1365

Table 1, continued

COVALENT BOND	POT./PROB.	BY SIM.	OTHER	TOTAL
PYRROLIDONE CARBOXYLIC ACID	65	157	457	679
HYDROXYLATION	65	159	371	595
METHYLATION	36	245	291	572
GAMMA-CARBOXYGLUTAMIC ACID	0	160	267	427
SULFATION	167	136	119	422
ADP-RIBOSYLATION	101	52	96	249
UBIQUITINATION	2	124	16	142
ALKYLATION	43	14	22	79
4-METHYLIDENE-IMIDAZOLE	0	63	1	64
DEAMIDATION	1	31	31	63
OXIDATIVE DEAMINATION	30	0	22	52
CITRULLINATION	1	40	6	47
D-ALANINE	0	1	38	39
D-ABU (AMINOBUTYRIC ACID)	0	2	36	38
OXIDATION (CYSTEINE)	0	36	1	37
FORMYLATION	9	0	27	36
SIALYLATION				34
INTERCHAIN CROSS-LINK	1	10	15	26
CHONDROITIN 4-SULFATE CROSSLINKING	10	9	3	22
2-AMINO-3-OXOPROPIONIC ACID	0	18	3	21
TYROSYLQUINONE (OR CROSS-LINKED TO)	0	21	0	21
FREE RADICAL	4	13	1	18
IODINATION	0	8	9	17
ASPIRIN-ACETYLATED SERINE	0	9	5	14
DHB (2,3-DIDEHYDROBUTYRINE)	0	1	11	12
DHA (2,3-DIDEHYDROALANINE)	0	0	11	11
SINGLE RESIDUE REMOVED	2	2	6	10
PROSTHETIC GROUPS (COVALENT COFACTORS)				
HEME (COVALENT)	58	389	586	1033
PYRIDOXAL PHOSPHATE	101	707	98	906
RETINAL CHROMOPHORE	0	142	67	209
PHOSPHOPANTETHEINE	45	117	18	180
SELENO-CYSTEINE	6	0	98	104
LIPOYL	25	60	16	101
PHYCOERYTHROBILIN CHROMOPHORE	0	0	97	97
PYRUVOYL	0	62	18	80
BIOTIN	0	41	8	49
TETRAPYRROLE CHROMOPHORE	5	24	12	41
DIPHTAMIDE	0	28	5	33
HYPUSINE	0	23	8	31
FAD (COVALENT)	0	22	6	28
FMN (COVALENT)	2	19	3	24
UMP (COVALENT)	0	12	6	18
TOPAQUINONE	0	11	6	17
TTQ	0	4	6	10

As suggested from the data in Table 1, only a small fraction of initiating and subsequently cleaved off methionines is annotated as such. Furthermore, it remains to be established whether all the annotated glycosylation sites (around 40000!) in SWISS-PROT are actually processed accordingly *in vivo* or just comply only with the very short PROSITE patterns (Falquet *et al.*, 2002). It is obvious that no currently available database can resemble a perfect state of annotations, especially regarding the status of experimental verification. While some annotations seem too optimistic (mainly those resulting from weak experimental or theoretical evidence), it is impossible, with the current literature retrieval systems, to keep a complete database up-to-date and cover all recent experimental results for existing entries.

Although each PTM group of entries in Table 1 might be considered as an initial learning set for the construction of a specific predictor, pure automatic retrieval of protein lists containing a specific modification for learning set creation is impossible since it requires extensive and time-consuming manual curation. Generally, the reliability of each PTM annotation can only be checked with a detailed analysis of the original scientific literature (Eisenhaber *et al.*, 1998; Nielsen *et al.*, 1999; Maurer-Stroh *et al.*, 2002a). Not all so-called experimentally verified examples are backed with unambiguous data in the scientific literature. Furthermore, SWISS-PROT sometimes supplies a commentary ("potential", "probable", "by similarity") indicating that the annotation at the time of entry production was inferred from non-experimental methods. On average, ~80% of the annotations are derived from some kind of theoretical consideration. Naturally, the question arises of how credible these annotations are. The answer depends on the type of modification being considered. There is a strong (and currently not satisfied) scientific need for 'gold standard' sequence sets for each PTM in which all instances are verified with the same identical, reliable technique. As we understand now, such 'gold standard' learning sets are not a trivial result but require tight interaction between interested theoretical and experimental researchers and, at least in some cases, the development of a new, medium-throughput technique for experimental verification of the PTM in many protein instances.

Is The Post-Translational Modification A Suitable Target For Predictor Construction?

A good predictor should be able to answer two questions:

1. Is the query protein a potential target for the PTM considered? Both the positive and negative answer need to be reliable and quantified for risk

assessment if the predictor is applicable for large-scale annotation tasks.

2. If it is, which are the sequence positions in the query protein that may be modified?

In this scheme, the capability for the considered PTM is thought to be encoded in one or several sequence segments of the query and this subsequence has to comply with requirements that are summarized in the recognition motif. In the simplest case, the motif can be expressed in the form of amino acid preferences at specific positions but, in the general case, it may be more degenerate and involve patterns of physical properties of amino acid side chains, possibly with correlation between motif positions.

Review of Biological Information on the PTM and Extraction of the Recognition Motif

Firstly, it is necessary to recall from the recent scientific literature the chemistry and the biology of the PTM under study. There should be experimental evidence that a recognition motif in the sequence of the protein does exist. It is helpful to know that protein factors or even enzymes assist in executing the PTM and knowledge of their binding properties might be of value. In the following, we will consider the standard case that an enzyme executing the PTM does exist and its specificity of binding a relevant portion of protein experiencing the PTM (also called substrate protein) contributes to the recognition motif.

Correlations between sequence patterns and biological features have to be established in advance before they are transferred to uncharacterised sequences in a later step. Unfortunately, the information pointing to sequence-function associations is often scattered in electronic databases and in the scientific literature and the extraction of the true protein sequence pattern from the heterogeneous data may become a time-consuming, independent scientific search of its own, as we have experienced with the representative cases of the GPI-lipid or myristoyl anchor sequence patterns (Eisenhaber *et al.*, 1998; Eisenhaber *et al.*, 1999; Maurer-Stroh *et al.*, 2002a; 2002b).

Learning Set Collection

It is necessary to collect sequence data on substrate proteins and specify a set of proteins that are experimentally verified or most likely to be modified with the PTM under investigation. Most often, SWISS-PROT is a good starting point in this effort but a lot of data on individual sequences remains scattered in the scientific literature and needs to be searched for. The latter is especially the case for mutational studies in which specific motif residues have been changed and tested for their influence on PTM efficiency (Eisenhaber *et al.*, 1999). Similarly, kinetic studies for the enzymes executing the PTM with model compounds are, as a rule, not reflected in databases (Rocque *et al.*, 1993; Towler *et al.*, 1988b). In the ideal case, this so-called learning-set samples the naturally occurring substrate sequence variability in the recognition motif region. The larger and better verified this sequence set is, the higher is the chance of getting a well-performing PTM predictor. If the literature indicates differing specificity of proteins having a role in the execution of the PTM with respect to experimentally analysed taxonomic ranges, the learning set should be split into taxon-specific classes (Towler *et al.*, 1988a). Such taxon specificity may be revealed also at a later time point when physical properties searched for in the recognition motif of learning set proteins are obviously related to their taxonomic classification (Eisenhaber *et al.*, 1998).

Finally, verified examples of sequences known to resist a PTM should be collected as a negative set. Sometimes, this helps to formulate typical sequence properties of non-modifiable proteins that improve their discrimination from modifiable ones (Eisenhaber *et al.*, 1999).

Redundancy Among the Sequences in the Learning Set

It could be desirable to reduce the number of sequences in the learning set by removing those with high sequence similarity to another sequence in the set. However, high sequence similarity within one group of proteins does not necessarily imply that all of them have the same capability for undergoing a given PTM. Sometimes one or two residue exchanges within the motif region suffice for abolishing the ability for the PTM considered. For example, there are three, sequentially closely related isoforms of the human folate receptor among which isoform α is a 100% GPI-lipid anchored, isoform β experiences this PTM only for about each second molecule copy, and isoform γ is not modified and wholly secreted. A good predictor must be able to distinguish among these cases (Eisenhaber *et al.*, 1999).

Thus, traditional techniques (Hobohm *et al.*, 1992) aimed at avoiding disproportional contribution of highly populated sequence subgroups can lead to loss of essential information on the recognition motif. As an alternative, there are a few methods that take into account both sequence- and position-specific weightings in profile extraction from alignments (such as PSIC (Sunyaev *et al.*, 1999)) and, in this way, try to overcome the problem of redundancy without loosing information from closely related, but different sequences.

Relationship of Recognition Motif and Substrate Protein 3D Structure

Despite of many years of intensive research, the prediction of structure and dynamics of a protein molecule as well as protein-ligand docking based on fundamental physical and chemical principles is still an unsolved scientific task in the general case (Eisenhaber *et al.*, 1995; Lomize *et al.*, 1999; Marchler-Bauer and Bryant, 1999). Therefore, theoretical protein science must settle for lesser goals such as the prediction of PTMs not encoded in structural features, because structural features cannot be currently predicted with sufficient reliability. Any prediction effort of these modifications can only be as good as the fold recognition and structure prediction itself. In this respect, the general prediction of the intra- and intermolecular disulfide bridge pattern appears not a good choice since it would be necessary to predict whether two cysteines are in a proper distance and orientation within 3D space. There are also other post-translational modifications that require distinct structural features within the substrate to be recognized by the processing machinery (e.g., for some palmitoylation, phosphorylation and glycosylation events). Neglecting the influence of three-dimensional features will result in loss of information and, thereby, reduce the reliability of predictions.

There are some post-translational modifications that occur in known non-globular regions of proteins such as unfolded N- or C-termini, polar low complexity regions and the like. As some examples, signal peptide cleavage (Nielsen *et al.*, 1999), a number of lipid modifications (GPI-lipid anchor attachment after C-terminal propeptide removal (Eisenhaber *et al.*, 1999), N-terminal N-myristoylation (Maurer-Stroh *et al.*, 2002b), farnesylation, geranylgeranylation) or separase digestion (Nasmyth *et al.*, 2000; Rao *et al.*, 2001) can be named. Since their prediction is apparently separated from other unsolved problems, these PTMs seem promising targets.

Biological Complexity of the Modification Process

The accessibility of post-translational modifications for predictions also depends on the modification processes themselves. How many different enzymes can facilitate the same or related modifications? How specific is the recognition between substrates and the processing enzymes? Is there a unique sequence signal?

In the case of myristoylation, the major part of known modified proteins is processed by only one enzyme (myristoylCoA:protein N-myristoyl-transferase) and information on the recognition sites, although quite weak, can be retrieved from the substrate sequences. In the case of prenylation, three different enzymes facilitate the modification (farnesyltransferase and geranylgeranyltransferases type I and II) and the sequence signal is less restricted. For palmitoylation, the processing enzymes are even still under discussion and, to a certain extent, a relatively nonspecific non-enzymatic catalysis is suspected. In the case of phosphorylation, although there are almost 5000 annotated sites in SWISS-PROT release 40, reliable prediction failed so far due to the enormous number of potentially involved enzymes (about 60 different kinases/phosphatases are known to modify the annotated proteins), due to the lack of a unique sequence signal, and also due to the involvement of tertiary structural features in a number of cases. Therefore, it is necessary to create a different learning set for every single modifying enzyme with unique substrate specificity. Maybe, the prediction of phosphorylation sites in protein substrates is promising if focused on individual, important, well-studied subclasses of kinases/phosphatases. The same strategy of concentrating on a subproblem appears suitable for proteolytic cleavage site predictions (proteolytic sites processed by about 30 different enzymes are annotated in SWISS-PROT release 40).

As the information on post-translational processing reactions increases, many more modifications will be accessible for prediction in the near future. On the other hand, this additional information may reveal exceptions to the known recognition motifs and many processes will turn out to be less specific or much more complex than previously thought.

Sequential Post-translational Modifications

When talking about post-translational processing, it has to be clear that proteins can undergo several different modifications that, furthermore, are often strongly interdependent and occur sequentially. For example, sialylation requires preceding glycosylation (Grabenhorst *et al.*, 1999). The diphtamide prosthetic group can be ADP-ribosylated (Marzouki *et al.*, 1991). Palmitoylation occurs more often in context with other signals for membrane attachment (such as myristoylation or prenylation) (Resh, 1999). In several cases of prenylation, the residues C-terminal to the lipid-modified amino acid are cleaved off and carboxymethylation follows (Sinensky, 2000). Attachment of a GPI-anchor to a protein is combined with the cleavage of its C-terminal propeptide after, typically, the protein has been exported to the endoplasmic reticulum with the signal leader mechanism (Kodukula *et al.*, 1995; Udenfriend and Kodukula, 1995). Even the small lantibiotics undergo an impressive number of post-translational modifications: dehydration of specific hydroxyl amino acids, enzymatic conversion of serine to D-alanine or of threonine to D-aminobutyric acid that then form thioether bonds to cysteines or build lysinoalanine bridges, oxidative decarboxylation of a C-terminal cysteine and cleavage of the propeptides (Sahl *et al.*, 1995).

Generally, maturation of proteins involves many steps of post-translational modifications, which in many cases are interdependent and, therefore, the sequence must contain signals to direct the whole cascade. In terms of predictability, these complementing features complicate the predictor structure but might bear additional information that can be relied on.

Construction Of A Prediction Function

Amino Acid Type Patterns of the PROSITE Type for the Recognition Motif Description

Conceptually, the simple description of re-occurring patterns in amino acid sequences with limited subsets of amino acid types at single positions (as in PROSITE (Falquet *et al.*, 2002)) is justified if either the set of all known sequences subdivides into disjunct clusters or if the pattern is tightly restricted and the correlation between sequence positions is marginal. However, the tight amino acid type restriction is rather the exception than the rule and, as a result, many amino acid type patterns for PTM motifs are often short, nonspecific, and produce a large number of false positive hits. It seems very difficult if not impossible, in this framework, to describe degenerated patterns of physico-chemical properties of amino acid side chains over sequentially long sequence segments with significant inter-positional correlations.

For relative comparison, we quantify the information carried by typical PROSITE patterns with an approach that uses the concept of Boltzmann's entropy. The probability P of occurrence of a pattern in a randomly generated amino acid sequence with pattern length N is

$$\boldsymbol{P} = \prod_{i=1}^{N} \boldsymbol{p}_i \qquad (1)$$

Here, $\boldsymbol{p}_i = \boldsymbol{n}_i / 20$ is the probability of matching the amino acid type at position $\boldsymbol{i}$ of the pattern if $\boldsymbol{n}_i$ is the number of allowed amino acid types at this position. Since PROSITE-type patterns do consider all positions as independent, the information content $\boldsymbol{R}$ can be calculated in bits as

$$\boldsymbol{R} = -\sum_{i=1}^{N} \mathrm{ld}(\boldsymbol{p}_i) \qquad (2)$$

Obviously, (1) more tightly restricted positions (smaller $\boldsymbol{p}_i$) or (2) more positions with restrictions (larger $\boldsymbol{N}$) imply more information and (3) a random pattern of any length ($\boldsymbol{p}_i = 1$) will always have zero information. The information content $\boldsymbol{R}$ can be considered as a measure for the predictive power of a PROSITE pattern with respect to the avoidance of false positive predictions and might, therefore, also be interpreted as reliability. Table 2 lists the PROSITE patterns of post-translational modifications and their $\boldsymbol{R}$ values. As expected, the short descriptors for phosphorylation, myristoylation and glycosylation sites obtain the lowest reliability in accordance to the common observation that those nonspecific patterns produce a frustratingly high number of false positive predictions.

As more experimentally characterized protein sequences are available, exceptions to generally accepted amino acid type motifs do occur, the patterns have to be relaxed and, thereby, become even more nonspecific. For example, this is the case for the glycine myristoylation motif (PROSITE pattern PS00008, see Table 2). This pattern finds about ~400,000 hits in SWISS-PROT (rel. 40), most of which are apparently wrong. Even worse, this motif produces false negative predictions for proteins whose myristoylation has been verified by experiment including ARF6_HUMAN (P26438), ARF6_CHICK (P26990), GCAP_BOVIN (P46065), HIA1_DICDI (P13231), HIA2_DICDI (P42526), HIPP_HUMAN (P41211), HIPP_RAT (P32076), NCAH_DROME (P42325), NECD_BOVIN (P29554), and NECX_APLCA (Q16982). The relaxed version in PROSITE notation reads as

G - {EDRHPFYW} - x(2) - [STAGCNDEF] - {P}

but the number of hits in SWISS-PROT using this pattern increases by about another 380,000.

Construction of a Discrimination Function

The concordance of a sequence segment of length $\boldsymbol{N}$ including sequence positions $\boldsymbol{n}, \boldsymbol{n}+1, \ldots, \boldsymbol{n}+\boldsymbol{N}$ in the query sequence $\boldsymbol{Q}$ can be quantified with a score function $\boldsymbol{S}$.

$$\boldsymbol{S}(\boldsymbol{n}, \boldsymbol{n}+1, \ldots, \boldsymbol{n}+\boldsymbol{N}) = \ln\left[\frac{\boldsymbol{P}(\boldsymbol{Q}(\boldsymbol{n}), \boldsymbol{Q}(\boldsymbol{n}+1), \ldots, \boldsymbol{Q}(\boldsymbol{n}+\boldsymbol{N}))}{\prod_{i=n}^{N} f(\boldsymbol{Q}(\boldsymbol{i}))}\right] \tag{3}$$

where $\boldsymbol{Q}(\boldsymbol{i})$ is the amino acid type at query sequence position $\boldsymbol{i}$ and $f(\boldsymbol{a})$ the expected occurrence of amino acid type $\boldsymbol{a}$ in protein sequences. $\boldsymbol{P}$ is the

Table 2. *R*-values (reliability) of PROSITE patterns from posttranslational modifications. The R-values have been calculated in accordance with equation (2) and are listed in bits. The relevant PROSITE entries are ordered with ascending information content.

PTM	**PROSITE**	**Pattern**	***R***
PHOSPHORYLATION BY PKC	PS00005	[ST]-x-[RK]	6.644
PHOSPHORYLATION BY CK2	PS00006	[ST]-x(2)-[DE]	6.644
PRENYLATION	PS00294	C-{DENQ}-[LIVM]-x>	6.966
G-MYRISTOYLATION	PS00008	G-{EDRKHPFYW}-x(2)-[STAGCN]-{P}	6.995
N-GLYCOSYLATION	PS00001	N-{P}-[ST]-{P}	7.792
PHOSPHORYLATION BY CAMP	PS00004	[RK](2)-x-[ST]	9.966
PHOSPHORYLATION BY TYROSINE-KINASES	PS00007	[RK]-x(2)-[DE]-x(3)-Y or [RK]-x(3)-[DE]-x(2)-Y	10.966
AMIDATION	PS00009	x-G-[RK]-[RK]	10.966
N-ACYL DIGLYCERIDE (PROKARYOTES ONLY)	PS00013	{DERK}(6)-[LIVMFWSTAG](2)-[LIVMFYSTAGCQ]-[AGS]-C	11.727
GLYCOSAMINOGLUCAN	PS00002	S-G-x-G	12.966
D/N-HYDROXYLATION	PS00010	C-x-[DN]-x(4)-[FY]-x-C-x-C	19.610

PROTEIN SELF-SPLICING (INTEIN)	PS00881	[DNEG]-x-[LIVFA]-[LIVMY]-[LVAST]-H-N-[STC]	19.703
E-CARBOXYLATION	PS00011	x(12)-E-x(3)-E-x-C-x(6)-[DEN]-x-[LIVMFY]-x(9)-[FYW]	20.177
PHOSPHOPANTETHEINE	PS00012	[DEQGSTALMKRH]-[LIVMFYSTAC]-[GNQ]-[LIVMFYAG]-[DNEKHS]-S- [LIVMST]-{PCFY}-[STAGCPQLIVMF]-[LIVMATN]-[DENQGTAKRHLM]-[LIVMWSTA]-[LIVGSTACR]-x(2)-[LIVMFA]	21.113
G-RADICAL	PS00850	[STIV]-x-R-[IVT]-[CSA]-G-Y-x-[GACV]	23.084
LIPOYL	PS00189	[GN]-x(2)-[LIVF]-x(5)-[LIVFC]-x(2)-[LIVFA]-x(3)-K-[STAIV]-[STAVQDN]-x(2)-[LIVMFS]-x(5)-[GCN]-x-[LIVMFY]	23.691
BIOTIN	PS00188	[GDN]-[DEQTR]-x-[LIVMFY]-x(2)-[LIVM]-x-[AIV]-M-K-[LVMAT]- x(3)-[LIVM]-x-[SAV]	27.236
PHOSPHORYLATION BY ATPASES E1, E2	PS00154	D-K-T-G-T-[LIVM]-[TI]	27.253
METHYLATION (PROKARYOTES ONLY, N-TERMINAL)	PS00409	[KRHEQSTAG]-G-[FYLIVM]-[ST]-[LT]-[LIVP]-E-[LIVMFWSTAG](14)	34.499

probability of the analysed query sequence segment to represent the recognition pattern having the length $\boldsymbol{N}$. Neither the functional form nor the parametric complexity of the function $\boldsymbol{P}$ are really known as well as the true motif length $\boldsymbol{N}$. Naturally, the function $\boldsymbol{P}$ can be evaluated with procedures for automated learning. Essentially, this means the generation of a complex regression between the learning set sequences and their capacity to undergo the post-translational modification. There is a variety of techniques such as neural networks (Hansen *et al.*, 1998; Blom *et al.*, 1999), Support Vector Machines (Hua and Sun, 2001), etc. If the learning set data are of sufficient quality and number and if the computational model is compatible with the physical realities, then the procedure will find the relationship without doubt. In the case of the detection of signal peptides and their cleavage sites, there is a wealth of information available regarding number of sequences and complexity of the motif. It is, therefore, accessible for automated learning methods (Nielsen *et al.*, 1999). Unfortunately, it is rarely possible to ensure these conditions. Although the number of known phosphorylated and glycosylated proteins is huge, nonspecificity of the motif results in a large number of false positive predictions for neuronal network approaches (Hansen *et al.*, 1998; Blom *et al.*, 1999). Following Table 1, the available sequence information for most PTMs is currently rather sparse. In the following text, we will explore in more detail such critical situations that apply to the majority of PTMs.

Let us consider a partition of P in the form

$$P(Q(n),\ldots,Q(n+N)) = \prod_{i} p_1(Q(i),i) \cdot \prod_{i,j} p_2(Q(i),Q(j),i,j) \cdot \prod_{i,j,k} p_3(Q(i),Q(j),Q(k),i,j,k) \cdot \ldots \tag{4}$$

where $p_1, p_2, p_3, \ldots$ are contributions that depend solely on a given single position, on two individual positions, on three motif positions and so on, respectively. Then, the total score can be calculated as

$$S(n,\ldots,n+N) = \sum_{i} ln\left(\frac{p_1(Q(i),i)}{f(Q(i))}\right) + \sum_{i,j} ln(p_2(\ldots,i,j)) + \ldots \tag{5}$$

The first term is the traditional profile expression that evaluates possible preferences for certain amino acid types at given motif positions (if $p_2 \equiv p_3 \equiv \ldots \equiv 1$). We want to emphasize that, for many PTMs, the profile term alone is insufficient for predictor construction since even single residue exchanges may convert a fully functional sequence into one that is no longer modified at all (Eisenhaber *et al.*, 1999).

The remaining terms assess the influence of multi-positional correlations in the total score. The motif target of a PTM can be degenerated in terms of amino acid preferences. This means that residues at a number of motif positions have to create a conserved environment with respect to physical conditions specific for the PTM and, as long as this constraint is fulfilled, the residues are free to vary their amino acid type. There are indeed such inter-positional correlations in the motif for GPI-lipid anchor attachment. For example, the residues ω-1...ω+2 are constrained in their side chain volume and backbone flexibility. Generally, tryptophane is a bad choice for position ω+1 but, if the other three are very small, then this residue may fit the motif (Kodukula *et al.*, 1993; Eisenhaber *et al.*, 1998). The condition of significant hydrophobicity in the C-terminal tail of the propeptide is another simple example (Udenfriend and Kodukula, 1995; Eisenhaber *et al.*, 1999).

Thus, the total score can be calculated as a sum of profile terms and physical property terms

$$S = S_{\text{profile}} + S_{\text{ppt}}. \tag{6}$$

This formulation includes both single-position penalties and terms evaluating correlations between several positions. This functional form leaves sufficient freedom to include analytical terms for conditions that have not been derived from positive examples in the learning set but also from analysis of enzymes executing PTMs or from sets of negative examples.

Parameterisation of the Score Function From the Biological Data

It is not trivial to define the functions $p_1, p_2, p_3, \ldots$ analytically. Again, one could use sophisticated optimisation tools such as genetic algorithms, taboo search, simulated annealing, etc. to parameterise the terms in S_{profile} and S_{ppt} and to achieve best discrimination for the current dataset and, thus, highest prediction accuracy. Since the experimental information on PTMs that can be relied on is not large and not very impressive regarding its quality, erroneous examples in the learning set may numerically bias the score function and reduce its applicability for yet uncharacterised sequences.

Simple approaches that try to impose explicit, plausible physical models onto the substrate protein recognition process appear more adequate since they are less prone to depend on deviating examples. Such a method has been described in Eisenhaber *et al.* (1999). The term p_1 has been approximated with a profile term complemented with position- and amino acid type-specific penalties. The remaining terms have been introduced as functions penalizing deviations from apparently conserved physical properties encoded in several residue positions or even whole subsegments of the motif. In this context, it is noted that the non-profile terms should generally only contribute negatively to the total score to exclude potential false positive predictions. Therefore, such prediction techniques might be more viewed as tools for rejecting clearly not appropriate target sequences in contrast to techniques optimised to find true hits.

For each recognition motif, amino acid type preferences and inter-positional correlations have to be computed from the learning set sequences or estimated from enzyme structures or from sets of negative examples. The comparison of relative amino acid type occurrences with amino acid indices (Tomii and Kanehisa, 1996) is helpful in uncovering the pattern of physical properties (Eisenhaber *et al.*, 1998). The determination of motif length is a tricky issue. Usually, one would expect a core motif region that is responsible for catalysis. There may be surrounding regions fitting the binding sites of the enzyme(s) subjected to less dramatic restrictions. There is often a third region that connects the site of modification with the rest of the substrate protein (an unstructured linker with preferentially flexible and hydrophilic residues). Sequence positions occupied primarily by hydrophobic residues that cannot be implicated in the binding process with the enzyme are usually signs of motif termination.

Evaluation Of False-Positive Prediction

A score alone does not tell how large the risk of making a false prediction is. A good predictor should translate the score into a probability of giving a false positive prediction calculated for a given query sequence.

First of all, it is necessary to generate a large sequence set of proven negative examples and to calculate their empirical score distribution function $F^*(S \leq S_{threshold})$. Typically, these sequences result in low scores. A suitable statistical model $F(S_{threshold})$ is then fitted to the empirical data. This analytical function can then be used to compute the probability P of random occurrence of a sequence segment fitting the motif requirements with a score $S \geq S_{threshold}$ via extrapolation into the region of high scores.

The statistical model depends on the PTM considered. If the recognition motif can maximally occur only at one place per query (for instance, if the motif is tied to the N-terminus as in the case of N-terminal N-myristoylation where the leading glycine is always the site of modification), one would expect the score being normally distributed as a first approximation and the probability of false positive prediction is calculated as

$$P(S \geq S_{threshold}) = \frac{1}{\sqrt{2\pi\sigma}} \int_{S_{threshold}}^{\infty} \exp - \left(\frac{S_{threshold} - \overline{S}}{\sigma} \right)^2 dS \tag{7}$$

In the second case, (1) the motif also occurs only once per query, but (2) the site is not bound to a fixed sequence position but may occupy a portion of the sequence, and (3) the length of this segment is fixed. For example, all three conditions are fulfilled in the case of the C-terminal GPI-lipid anchor attachment site. Then the statistical model incorporated in the extreme-value distribution appears more applicable.

$$P(S \geq S_{threshold}) = 1 - exp\left(-e^{-f(S_{threshold})}\right) \tag{8}$$

with

$$f(x) = \lambda(x - u) \tag{9}$$

where λ and u are parameters. If a score S is normally distributed, then the probability of a score S to be larger than a threshold $S_{threshold}$ is described by an extreme-value distribution; i.e., the local maximum of S is extreme-value distributed.

If the motif may occur one or several times in one sequence, or if the sequence segment containing the potential site of modification has variable length (for example, in the case of transmembrane prediction), then the probability calculated in (8) needs to be corrected with the query sequence length divided by the motif length.

The considerations above are correct if all terms of the total score S in equations (5) and (6) are independent of each other and normally distributed since, then, S is also normally distributed. This analytical form is, as a rule, difficult or impossible to achieve. Nevertheless very often, this first approximation works fine (Eisenhaber *et al.*, 2001) but a more accurate estimate of the probability might be needed for some applications. Therefore, generalized analytical functions have to be introduced that reasonably take these term-term interactions into account (Eisenhaber *et al.*, 2001). The functional form of (8) in combination with

$$f(x) = \sum_i \lambda_i (x - u_i)^{\gamma_i} \tag{10}$$

(the generalized extreme-value distribution) has been a practical solution in the case of the GPI-lipid anchor attachment predictor.

Conclusions

The prediction of post-translational modifications will be one of the major fields in biomolecular sequence analysis in the near future. Although existing methodological approaches based on the available biological data can be helpful in the construction of efficient predictors, a number of major breakthroughs will critically depend on the supply of 'gold standard' learning sets. Theoretical approaches can be helpful in the selection of sequence lists for experimental test and, thereby, dramatically reduce the unavoidable experimental load.

Acknowledgements

The beginning of our research on post-translational modifications would have been unthinkable without the input and encouragement from Peer Bork (EMBL, Heidelberg) and Jens G. Reich (MDC, Berlin). The authors are also grateful for generous financial support from Boehringer Ingelheim. This project has been partly funded by the Austrian National Bank (Österreichische Nationalbank) and by the Fonds zur Förderung der wissenschaftlichen Forschung Österreichs (FWF P15037).

References

Bairoch, A. and Apweiler, R. 2000. The SWISS-PROT protein sequence database and its supplement TrEMBL in 2000. Nucleic Acids Res. 28: 45-48.

Blom, N., Gammeltoft, S., and Brunak, S. 1999. Sequence and structure-based prediction of eukaryotic protein phosphorylation sites. J. Mol. Biol. 294: 1351-1362.

Bork, P., Dandekar, T., Diaz-Lazcoz, Y., Eisenhaber, F., Huynen, M., and Yuan, Y. 1998. Predicting function: from genes to genomes and back. J. Mol. Biol. 283: 707-725.

Eisenhaber, B., Bork, P., and Eisenhaber, F. 1998. Sequence properties of GPI-anchored proteins near the omega-site: constraints for the polypeptide binding site of the putative transamidase. Protein Eng. 11: 1155-1161.

Eisenhaber, B., Bork, P., and Eisenhaber, F. 1999. Prediction of potential GPI-modification sites in proprotein sequences. J. Mol. Biol. 292: 741-758.

Eisenhaber, B., Bork, P., and Eisenhaber, F. 2001. Post-translational GPI lipid anchor modification of proteins in kingdoms of life: analysis of protein sequence data from complete genomes. Protein Eng. 14: 17-25.

Eisenhaber, B., Bork, P., Yuan, Y., Loffler, G., and Eisenhaber, F. 2000. Automated annotation of GPI anchor sites: case study *C. elegans*. Trends Biochem. Sci. 25: 340-341.

Eisenhaber, F. and Bork, P. 1998. Sequence and structure of proteins. In: Recombinant Proteins, Monoclonal Antibodies and Theraeutic Genes. Schomburg, D., ed. Volume 5 of Biotechnology. 2 ed. Rehm, H.-J. and Reed, G., eds. Wiley-VCH. p. 43-86

Eisenhaber, F., Persson, B., and Argos, P. 1995. Protein structure prediction: recognition of primary, secondary, and tertiary structural features from amino acid sequence. Crit. Rev. Biochem. Mol. Biol. 30: 1-94.

Falquet, L., Pagni, M., Bucher, P., Hulo, N., Sigrist, C.J., Hofmann, K., and Bairoch, A. 2002. The PROSITE database, its status in 2002. Nucleic Acids Res. 30: 235-238.

Grabenhorst, E., Schlenke, P., Pohl, S., Nimtz, M., and Conradt, H.S. 1999. Genetic engineering of recombinant glycoproteins and the glycosylation pathway in mammalian host cells. Glycoconj. J. 16: 81-97.

Hansen, J.E., Lund, O., Tolstrup, N., Gooley, A.A., Williams, K.L., and Brunak, S. 1998. NetOglyc: prediction of mucin type O-glycosylation sites based on sequence context and surface accessibility. Glycoconj. J. 15: 115-130.

Harding, J.J. 1985. Nonenzymatic covalent post-translational modification of proteins *in vivo*. Adv.Protein Chem. 37: 247-334

Hobohm, U., Scharf, M., Schneider, R., and Sander, C. 1992. Selection of representative protein data sets. Protein Sci. 1: 409-417.

Hua, S. and Sun, Z. 2001. Support vector machine approach for protein subcellular localization prediction. Bioinformatics. 17: 721-728.

Jaffrey, S.R., Erdjument-Bromage, H., Ferris, C.D., Tempst, P., and Snyder, S.H. 2001. Protein S-nitrosylation: a physiological signal for neuronal nitric oxide. Nat. Cell Biol. 3: 193-197.

Kodukula, K., Gerber, L.D., Amthauer, R., Brink, L., and Udenfriend, S. 1993. Biosynthesis of glycosylphosphatidylinositol (GPI)-anchored membrane proteins in intact cells: specific amino acid requirements adjacent to the site of cleavage and GPI attachment. J. Cell Biol. 120: 657-664.

Kodukula, K., Maxwell, S.E., and Udenfriend, S. 1995. Processing of nascent proteins to glycosylphosphatidylinositol-anchored forms in cell-free systems. Methods Enzymol. 250: 536-547.

Lee, A.T. and Cerami, A. 1992. Role of glycation in aging. Ann. N. Y. Acad. Sci. 663: 63-70.

Lomize, A.L., Pogozheva, I.D., and Mosberg, H.I. 1999. Prediction of protein structure: The problem of fold multiplicity. Proteins 37: 199-203.

Marchler-Bauer, A. and Bryant, S.H. 1999. A measure of progress in fold recognition? Proteins Suppl. 3: 218-225.

Marzouki, A., Sontag, B., Lavergne, J.P., Vidonne, C., Reboud, J.P., and Reboud, A.M. 1991. Effect of ADP-ribosylation and phosphorylation on the interaction of elongation factor 2 with guanylic nucleotides. Biochimie 73: 1151-1156.

Maurer-Stroh, S., Eisenhaber, B., and Eisenhaber, F. 2002a. N-terminal N-myristoylation of proteins: Refinement of the sequence motif and its taxon-specific differences. J. Mol. Biol. 317: 525-542.

Maurer-Stroh, S., Eisenhaber, B., and Eisenhaber, F. 2002b. N-terminal N-myristoylation of proteins: Prediction of substrate proteins from amino acid sequence. J. Mol. Biol. 317: 543-559.

Nasmyth, K., Peters, J.M., and Uhlmann, F. 2000. Splitting the chromosome: cutting the ties that bind sister chromatids. Science 288: 1379-1385.

Nielsen, H., Brunak, S., and von Heijne, G. 1999. Machine learning approaches for the prediction of signal peptides and other protein sorting signals. Protein Eng. 12: 3-9.

Rao, H., Uhlmann, F., Nasmyth, K., and Varshavsky, A. 2001. Degradation of a cohesin subunit by the N-end rule pathway is essential for chromosome stability. Nature 410: 955-959.

Rattan, S.I., Derventzi, A., and Clark, B.F. 1992. Protein synthesis, post-translational modifications, and aging. Ann. N. Y. Acad. Sci. 663: 48-62.

Resh, M.D. 1999. Fatty acylation of proteins: new insights into membrane targeting of myristoylated and palmitoylated proteins. Biochim. Biophys. Acta 1451: 1-16.

Robinson, N.E. and Robinson, A.B. 2001. Prediction of protein deamidation rates from primary and three- dimensional structure. Proc. Natl. Acad. Sci. USA 98: 4367-4372.

Rocque, W.J., McWherter, C.A., Wood, D.C., and Gordon, J.I. 1993. A comparative analysis of the kinetic mechanism and peptide substrate specificity of human and *Saccharomyces cerevisiae* myristoyl-CoA:protein N-myristoyltransferase. J. Biol. Chem. 268: 9964-9971.

Sahl, H.G., Jack, R.W., and Bierbaum, G. 1995. Biosynthesis and biological activities of lantibiotics with unique post- translational modifications. Eur. J. Biochem. 230: 827-853.

Sinensky, M. 2000. Recent advances in the study of prenylated proteins. Biochim. Biophys. Acta 1484: 93-106.

Sunyaev, S.R., Eisenhaber, F., Rodchenkov, I.V., Eisenhaber, B., Tumanyan, V.G., and Kuznetsov, E.N. 1999. PSIC: profile extraction from sequence alignments with position- specific counts of independent observations. Protein Eng. 12: 387-394.

Tomii, K. and Kanehisa, M. 1996. Analysis of amino acid indices and mutation matrices for sequence comparison and structure prediction of proteins. Protein Eng. 9: 27-36.

Towler, D.A., Adams, S.P., Eubanks, S.R., Towery, D.S., Jackson-Machelski, E., Glaser, L., and Gordon, J.I. 1988a. Myristoyl CoA:protein N-myristoyltransferase activities from rat liver and yeast possess overlapping yet distinct peptide substrate specificities. J. Biol. Chem. 263: 1784-1790.

Towler, D.A., Gordon, J.I., Adams, S.P., and Glaser, L. 1988b. The biology and enzymology of eukaryotic protein acylation. Annu. Rev. Biochem. 57: 69-99.

Udenfriend, S. and Kodukula, K. 1995. How glycosylphosphatidylinositol-anchored membrane proteins are made. Annu. Rev. Biochem. 64: 563-591.

Uy, R. and Wold, F. 1977. Post-translational covalent modification of proteins. Science 198: 890-896.

From: *Bioinformatics and Genomes: Current Perspectives*
Edited by: Miguel A. Andrade

Chapter 6

Automatic Genome Annotation and the Status of Sequence Databases

Miguel A. Andrade

Abstract

The characterisation of the genes predicted from a completely sequenced genome gives a detailed view of an organism. However, most of the proteins deduced from a genome project have never been seen before, and cannot be biochemically characterised on a short time scale. Therefore, computational methods are used to annotate the function of new protein sequences based on similar sequences already annotated in the sequence databases. However, the complex relation between function and sequence makes this difficult and erroneous annotation can happen. New errors can be produced when a wrongly annotated entry is used to annotate a new protein. In this chapter, I discuss the problems of function inference by protein sequence similarity, considering also the status of databases in relation to annotation errors. Finally, I evaluate the inter-relation between databases and the annotation process by similarity (using an automatic system) and how this situation is likely to evolve.

INTRODUCTION

The initial result of a genome project is a complete genomic sequence. Ideally, it should be possible to infer from a genome the complete set of proteins present in an organism. Discovering the functions of all those proteins is very important for a better understanding of how the corresponding organism works and how it relates to the environment. However, it is unfeasible to experimentally characterise the usually thousands of hypothetical proteins derived from a new genome. While awaiting experimental characterisation, the annotation of a genome using computational methods aims to convey as much experimental information as possible to hypothetical proteins from similar sequences in the databases. As a result, experiments can be targeted towards interesting candidates.

Protein annotation by sequence similarity starts with a simple assumption: a given protein will have the same function as another protein provided that they share a certain degree of sequence similarity. Unfortunately, this elementary approach has some limitations (see for example, Bork and Bairoch, 1996; Devos and Valencia, 2000; Ponting *et al.*, 2000; Hegyi and Gernstein, 2001). Two similar proteins can have different substrate specificity (e.g. glucose transporters are similar to maltose transporters). Two proteins sharing a domain can have different functions (e.g. there are protein tyrosine kinases and protein tyrosine phosphatases that have an SH2 domain).

Accordingly, genome annotation by homology is error-prone as can be seen in a multitude of protein annotations within genome projects (discussed, for example in Galperin and Koonin, 1998; Devos and Valencia, 2001). Some of these erroneous annotations are deposited in the databases, constituting a possible threat for further annotation by similarity. Obviously, this kind of mistakes is due to the haste with which a genome of thousands of genes is usually annotated. Many of the thousands of gene annotation actions will create particular problems, requiring knowledge of numerous different aspects of the molecular biology relating to an organism that is often poorly understood anyway. Functional inference using sequence similarity may be helpful in such a case as a starting point, but then there might not be time left for correcting possible mistakes.

With this problem in mind, automatic tools for the support of genome annotation have been developed such as GeneQuiz, which originated in the pre-genomic era (Scharf *et al.*, 1994) and evolved as a result of it (Andrade *et al.*, 1999), Magpie (Gaasterland and Sensen, 1996), or PEDANT (Frischman *et al.*, 2001). In particular, GeneQuiz is available as a web server for analysing sequences (Hoersch *et al.*, 2000) and can perform whole genome analyses (Iliopoulos *et al.*, 2000). The main idea of such a system is to perform automatically repetitive steps such as the preparation of the databases to be used for sequence comparison, running several complementary methods for protein analysis, displaying the results, and the production of tentative suggestions for functional annotation. Basically, GeneQuiz is composed of a series of modules: GQUpdate keeps up-to-date a series of databases that are used for sequence comparison. GQSearch takes a protein sequence as input, and invokes the sequence comparison methods BLAST (Altschul *et al.*, 1990) and FASTA (Pearson, 1996), as well as other methods for protein analysis (secondary structure prediction, domain detection, and delineation of transmembrane segments, low complexity regions and coiled coil regions). GQReason then creates a summary of the homologues and evaluates their potential as source for transferring a functional description to the query (in four categories: clear, tentative, marginal and negligible). The latter evaluation is based on the sequence similarity values given by BLAST and FASTA, the database of origin of the sequence, and the quality of the annotation. The best annotation according to this procedure is then suggested for the query, and reported together with its associated reliability category.

In the next sections, I will discuss the main causes of error in functional annotation by similarity. I will also propose possible solutions that can be helpful either for improving automatic systems for large-scale annotation or for the researcher doing everyday analysis on a single sequence.

The Problems Transferring Function Between Similar Proteins

Let us consider a typical example of annotation error deposited in a database, which illustrates several factors that might lead to an incorrect functional annotation.

An Example of Erroneous Annotation

The entry Q9V0Y0 of the SpTrembl database (Bairoch and Apweiler, 2000) is currently annotated as "nodulation protein" (October 2001). This annotation comes from the complete genome sequencing analysis and annotation of the archeobacteria *Pyrococcus abyssi* (see Figure 1). On the one hand, *Pyrococcus abyssi* was isolated from samples close to a hot water spring situated 3,500 meters deep in the Pacific. This organism grows optimally at 103 °C and 200 atmospheres pressure. On the other hand, nodulation is a process by which some bacteria produce nodules on the roots of legumes hosting them. It is extremely unlikely that a *P. abyssi* sees a legume in its life, and therefore the annotation does not make sense.

```
ID   Q9V0Y0      PRELIMINARY;      PRT;   440 AA.
AC   Q9V0Y0;
DT   01-MAY-2000 (TrEMBLrel. 13, Created)
DT   01-MAY-2000 (TrEMBLrel. 13, Last sequence update)
DT   01-JUN-2001 (TrEMBLrel. 17, Last annotation update)
DE   NODULATION PROTEIN NFED.
GN   PAB1932.
OS   Pyrococcus abyssi.
OC   Archaea; Euryarchaeota; Thermococcales; Thermococcaceae; Pyrococcus.
OX   NCBI_TaxID=29292;
RN   [1]
RP   SEQUENCE FROM N.A.
RC   STRAIN=ORSAY;
RA   Heilig R.;
RT   "Pyrococcus abyssi genome sequence: insights into archaeal chromosome
RT   structure and evolution.";
RL   Submitted (JUL-1999) to the EMBL/GenBank/DDBJ databases.
DR   EMBL; AJ248285; CAB49572.1; -.
DR   InterPro; IPR002810; DUF107.
DR   Pfam; PF01957; DUF107; 1.
KW   Complete proteome.
SQ   SEQUENCE   440 AA;  47956 MW;  775239072B1A04E0 CRC64;
     MKKIILPLLV FAFIASPVLA GNVVYVAQIK GQITSYTYDQ FDRYITIAEE NNAEAIIIEF
     DTPGGRADAM MNIIQRIQQS KVPVIIYVYP PGATAASAGT YIALGSHLIA MAPGTSIGAC
     RPILGYSQNG SIIEAPPKIT NHFIAYIKSL AQESGRNETI AEEFITKDLS LTPEEALKYG
     VIEVIARDVN ELLGKANGMK TKLPVNGRYV TLNFTNVDVK YLSPSLKDKL IMYITDPNVA
     YLLLTLGIWA LIIGFLTPGW HVPETVGAIM VILAIIGFGY FGYNSAGILL IIIAMLFFVA
     EALTPTFGLF TVAGLITFII GGILLFGGGE EYLIKREVFS QLRILIITVG VILAAFFAFG
     MAAVIRAHRR KASTGREEMI GLTGTVVEEL NPEGMVKVRG ELWRARSKFG EKIEKGERIK
     VVDIEGLTLI VVREGKGGER
//
```

Figure 1. Example of wrong annotation. The database entry of a protein sequence (in SpTrembl; Bairoch and Apweiler, 2000) from the archeobacterium *Pyrococcus abyssi* is displayed with three important fields in boldface. The uppermost refers to the description of the protein. In this case indicates a function: "nodulation protein NFED". The middle marked field points to the species from which the protein is originated: *Pyrococcus abyssi.* The last marked field indicates a literature pointer with information about the protein. In this case, notice that the reference corresponds to a genome project. It is very likely that the protein sequence is the result of the conceptual translation of an open reading frame derived from the genomic sequence. The annotation must be wrong because "nodulation" is a process that is incompatible with the properties of *P. abyssi* (see text).

The origin of this annotation can be easily tracked. If we compare the original sequence against the databases of protein sequences, the nearest homologue with an experimental characterisation (as of October 2001, the only one) is a protein from *Rhizobium etli*. The *R. etli* database entry details the authors who submitted the sequence to the database, though there is no link to any reviewed paper. However, it is possible to retrieve a related paper from the MEDLINE literature database by the same authors (Borthakur and Gao, 1996) where it is shown that the protein was implicated in "nodulation competitiveness" (a somewhat vague indication of function). It is obvious that the annotator of the *P. abyssi* sequence just took the word "nodulation" without further consideration, and later this annotation was taken by the database curators.

Other important information is also missing in the *P. abyssi* entry: the domain organisation of the sequence that might influence function prediction. This sequence belongs to a widespread prokaryotic family (with most sequences also originating from sequencing projects with incorrect annotation). The length disparity of the *P. abyssi* homologues indicates that it might contain two domains: while the *R. etli* homologue has the same length as the *P. abyssi* sequence, there is a homologue from another archaean, *M. jannaschii*, which is considerably shorter. Indeed, the whole family presents several short and long members (even both existing in single organisms).

In summary, the members of the whole prokaryotic family including the *P. abyssi* sequence should all be re-annotated and the presence of either one or two domains stated. The published function should not be adopted by these sequences because the experimental evidence "implicated in nodulation competitiveness" is too general, and also because there is no evidence linking this general function to either of the domains.

Through this example, we have seen some of the possible factors that can result in mistakes in functional annotation by similarity. If the annotation from a source sequence in the database is transferred to a query sequence, the transference could be wrong if query and source have different domain organisation, different functional specificity, or if the source is already wrongly annotated. Now I will discuss in more detail each of these three possibilities.

Different Domain Organisation

The transfer of function from one protein to another should be done only when the proteins match over their total lengths, without any large sequence fragments left out in the alignment. If some fragments do not align, one has to be very careful in checking whether the function that is transferred really corresponds to the part that is common. The use of domain databases and of methods of domain identification can be of great help at this point.

In any case, if a query sequence contains a domain that is not matched, this should be noted. And, secondly, if the protein used as source of the annotation contains an unmatched domain, the annotation should just state the presence of the domain(s) of the query without extrapolating a total functional identity.

Wrong Assesment of Functional Specificity

This problem applies when the query is similar to proteins with a roughly identical function but with subtle variations (e.g. dehydrogenases that act on different substrates). Insight can be gained by aligning the members of the family and making a phylogenetic analysis of the family as accurate as possible. If the family is organised into a tree that correlates nicely with the functions described (or derived) for the proteins, then one can draw conclusions for the query. If the query falls into a clear group, then one can suggest the function of the branch. If the function of the branch is heterogeneous, or the query falls in between branches, one should choose carefully a less specific description matching the closer branches (e.g. "alcohol dehydrogenase"), or plainly give up and use a generic description of the whole family (e.g. "dehydrogenase").

In the case of full genome annotation, once a first draft annotation is produced, one can take advantage of the fact that the existence of one protein might need the existence of another. If a protein produces a component, this has to be used or transported by another protein. Also the presence of one of the components of a complex points to the existence of all the other components of the complex. And so on. This can be of great help when deciding between alternative tentative functions for one protein.

For example, if all the proteins of a linear metabolic pathway are present except for one that performs an intermediate step, one can expect that the missing activity has to be performed by some undetected protein. It can happen that the missing protein was wrongly assumed to perform some other activity. It is also possible that a different enzyme (non-homologous to the one known to perform the missing activity in another organisms) is doing the task. This is known as gene orthologous gene displacement (Koonin *et al.*, 1996). In such a case, sequence similarity analysis cannot help much. In the case of prokaryotic organisms, a last resource would be to use the fact that proteins close in the genome tend to be functionally related. Given a protein, the neighbourhood of its homologs in complete prokaryotic genomes can be analysed using the STRING server (Snel *et al.*, 2000). These supporting strategies have been used, for example, for the re-annotation of the *Mycoplasma pneumoniae* genome (Dandekar, *et al.*, 2000).

Errors in the Database

If the protein from which the annotation is transferred is wrongly annotated, then it is likely that the annotation will also be wrong. The errors in the database belong to the aforementioned two categories: errors in the assessment of the domain organisation, and in the specificity of function assigned. They can be corrected as described above. A way of mitigating further error propagation is to check the origin of the annotation. Annotations derived by an experiment should be given maximum reliability (although errors are also possible at this level). An additional error that has to be considered is spelling mistakes. They require just being careful enough to read the annotation considered.

Testing Function Transference

An additional complication with the errors in the database can come from its extent. In assigning a function to a query protein by similarity, in principle one single transference has to be checked. But in order to avoid the possibility of an error in the database sequence that is used as source of the transference, one has also to check the way that source entry was annotated. If the distribution of functions associated to the homologues is complicated, this might imply checking several annotations, aligning many sequences, making a phylogenetic tree, searching for experimentally characterised proteins, examining the corresponding literature, and the like; altogether leading to an effort that amounts to correcting the database itself. This cannot be done for the analysis of a complete genome. In practise, it is not always possible to produce good quality annotations, so that wrong annotations are bound to enter the databases, thus increasing the fraction of wrongly annotated sequences and making future annotation more difficult.

How bad is the situation of the current databases regarding functional annotation? And to what extent are the genome projects responsible for this situation? How does it affect transference of function by sequence similarity? In order to analyse these points I tested a number of protein database entries, and the transfer of function from protein sequence databases to new proteins of unknown function using GeneQuiz.

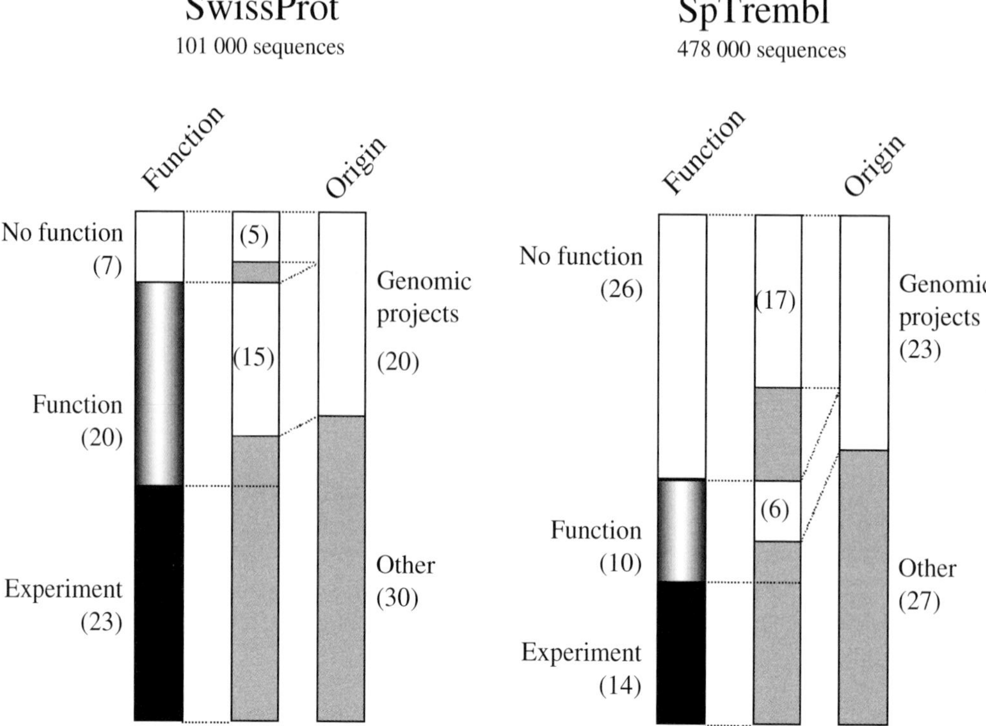

Figure 2. Estimation of the current status (October 2001) of the functional annotation in two protein databases (SwissProt, SpTrembl; Bairoch and Apweiler, 2000). Each database was tested with 50 entries randomly selected (http://kr.expasy.org/sprot/get-random-entry.html). The quality of the functional annotation is described as: "experiment", the entry is linked to a paper that reports an experiment on the function; "function", the protein has a functional annotation but it has not been experimentally checked; "no function", the protein lacks functional annotation. Numbers is parentheses indicate the number of entries observed in each category. The rightmost column for each database indicates the origin of the sequences that can be a "genome project" or other. The middle column discloses the origin of the sequences for each of the functional categories. SwissProt and SpTrembl are produced in a coordinated way in EBI (Cambridge, UK) and the SiB (Geneve, Switzerland). New sequences are introduced in SpTrembl, and eventually they are transferred to SwissProt where they go through a stricter curation process. The sequences from genome projects during the second half of the 1990s have made it through SpTrembl to Swissprot, and today there is no large difference in genomic origin between those databases (with almost a 1/2 of sequences coming from genome projects). This fraction is likely to tilt further as the production of sequences from genome projects keeps increasing. The largest difference between databases, which reflects the superior level of curation of SwissProt over SpTrembl, regards the more abundant experimental fraction in SwissProt, and especially, the larger fraction of functionally annotated sequences (irrespective of their origin). If both databases are taken together, the relation 5:1 in number of sequences makes the total to look more than SpTrembl. In general, the complete database has therefore a large fraction of sequences without any annotation, and a considerable fraction of sequences annotated by similarity (about a third), many of them from genome projects, with implications for the introduction of errors in databases as we will see.

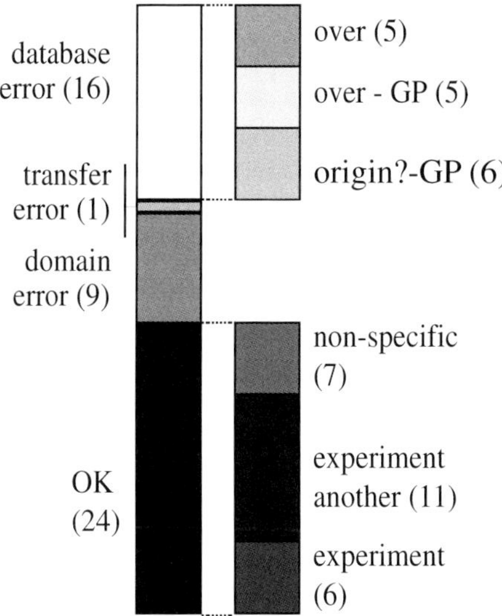

Figure 3. Analysis of automatic annotation. The GeneQuiz run of September 2001 on the genome of *Sulfolobus tokadaii* was used as a test to evaluate the performance of annotation by similarity, and the relation of the status of the databases to this performance. This archaeon is closely related to another two archaea, *Sulfolobus solfataricum* and *Aeropyrum pernix*. Experimental information is very scarce for these three organisms. However, the system produced 1,500 clear predictions (out of the 2,825 predicted proteins) using similarity to annotated proteins in the databases. 50 randomly selected predictions were manually analysed. Left column indicates the accuracy of the annotation: OK, for correct annotations, and the other categories for possible errors. Numbers indicate the number of entries in each category. Almost half of the predictions were all right. Bottom-right column: origin of correctly transferred annotations: experimentally characterised proteins, proteins annotated via similarity, and proteins annotated with correct non-specific annotations. This latter category is remarkable in the sense that it reflects a successful effort in annotating proteins with a function that we could call *synthetic*: such annotations are not correct in strict sense because they are incomplete (for example, "sugar transporter"; it is not explaining which sugar or sugars are transported), but they are in fact optimal in the absence of phylogenetic information leading to a more concrete conclusion. It is clear that an ontologic definition of function should be of great help for cases like these (Ashburner *et al.*, 2000). The system failed 1/2 of the cases in various degrees of seriousness (discussed in Iliopoulos *et al.*, submitted). In one case, the transfer of function was done from the incorrect sequence. In nine more cases, there was a different domain organization in the query and in the database sequence used for transferring the function. Seven of those cases could be detected by using databases of domains (Pfam, Bateman *et al.*, 2000; SMART, Schultz *et al.*, 2000). Interestingly, the remaining 1/3 of erroneous annotations was due to transference from proteins erroneously annotated in the database. The top-right column details the origin of those errors: over-predictions of function in sequences of non-genomic origin (over) or genomic origin (over-GP), or annotations of unknown origin from genomic projects (origin?-GP). Most of the errors were over-predictions of function (for example, predicting an RNA ATP-dependent helicase when the family did not show conclusively whether the substrate was DNA or RNA), and most of them originated in sequences from genome projects. No mistake was done when the sequence had similar length and was experimentally characterised. The relation of length to transfer accuracy is analysed in the next figure.

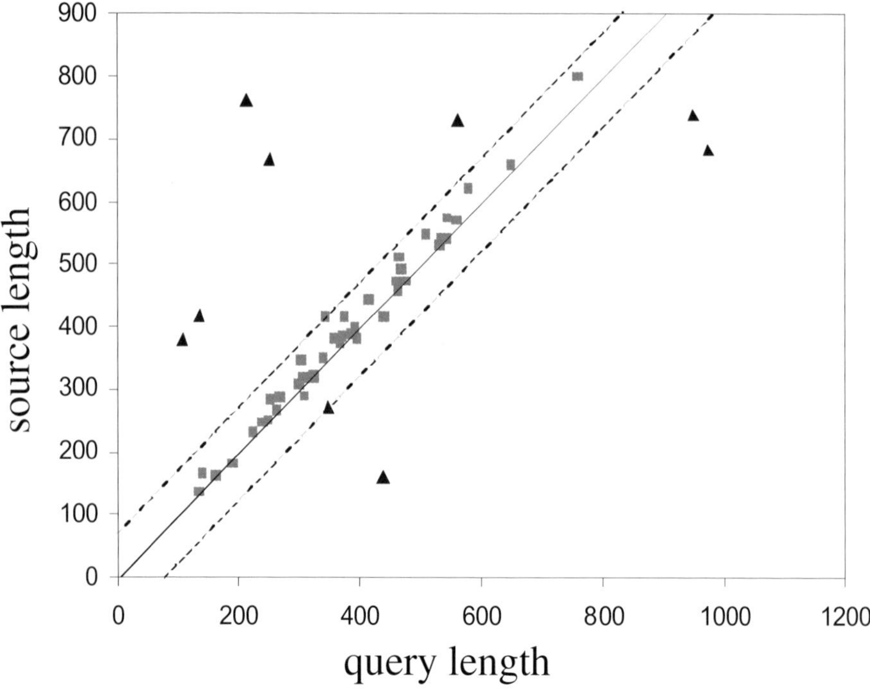

Figure 4. Relation between the difference in length between compared sequences and the accuracy of the transference for the 50 cases for which automatic annotation was predicted, producing either a correct (squares) or an incorrect transfer (triangles). The nine erroneous cases are all above 60 amino acids of difference. Other errors observed do not depend on length.

I have estimated the status of functional annotation in protein databases using a random sampling of sequences from two of them (see Figure 2). The most apparent feature is the wealth of sequences originated from genome sequencing projects (currently close to 50%). Functional annotation of the genomic part of the database is scarce (1/3) in comparison to that of the rest (2/3), resulting in the annotation of half of the sequences. Of the annotated fraction, a further half has some associated experimental evidence (almost non-existent in the case of genomic sequences).

In order to analyse how functional transference via similarity works today and how its performance is affected by the status of the databases, I tested the transfer of function to new proteins of unknown function. For this I used a recent automatic analysis of GeneQuiz (September 2001) on the genome of the archaea *Sulfolobus tokodaii*. The selected test set of genes was taken from the "clear" associations (those with the highest reliability level). Two kinds of homologues were considered: the closest with meaningful functional annotation (the one taken, in principle, by GeneQuiz), and the closest with experimental information. For each case I considered possible errors in the transfer (domain, specificity, transfer from the wrong protein, or errors in the original database), and the role of genome projects in the introduction of database errors (see Figure 3).

The automatic annotation was correct in half of the cases. Mistakes were mostly due to errors in database entries, most of them corresponding to genome projects. Function over-prediction (assignment of a specific function without enough phylogenetic evidence) was the most common mistake. This indicates that there is certainly a necessity of reassessing databases regarding the specificity of function annotations of many sequences inferred by similarity. This has been done already for a significant amount of sequences as implicated by the high fraction of automatic results that have a reasonable non-specific annotation taken from the database.

The domain problem could be detected from the difference in the length of the query and the database sequence (see Figure 4). In the cases analysed here, all pairs query-source sequence where the length difference was of more than 60 amino acids had a different domain organisation. Obviously, such a threshold should be increased when considering longer proteins.

Conclusion

Today, gene annotation using sequence similarity could produce an increase in the fraction of erroneously annotated sequences in databases. Genome projects actually avoid this by submitting relatively few annotations which, however, reduces the fraction of functionally annotated proteins in databases. Nonetheless, this conservative action produces less damage in the databases (see Figure 5). As we have seen, protein annotation via similarity is too dangerous to be used without care. The good news is that most of the problems of gene annotation via similarity are not due to the annotation method itself, but to the transference of errors already present in the databases.

In fact, there is an effect that favours the simple approach of taking just one sequence from the database and transferring its annotation to a query. As many more sequences enter the databases, it becomes more likely that the closest homologue with an annotated function has the same distribution of domains and function (which validates the functional transference). Therefore, it is probably reasonable to keep the simple approach as basic step, while investing most efforts in improving the accuracy of the databases used for annotation.

Problems of over-prediction of function of poorly characterised protein families should be detected and corrected by providing less specific annotations. For this, family analysis, including the alignment of the members

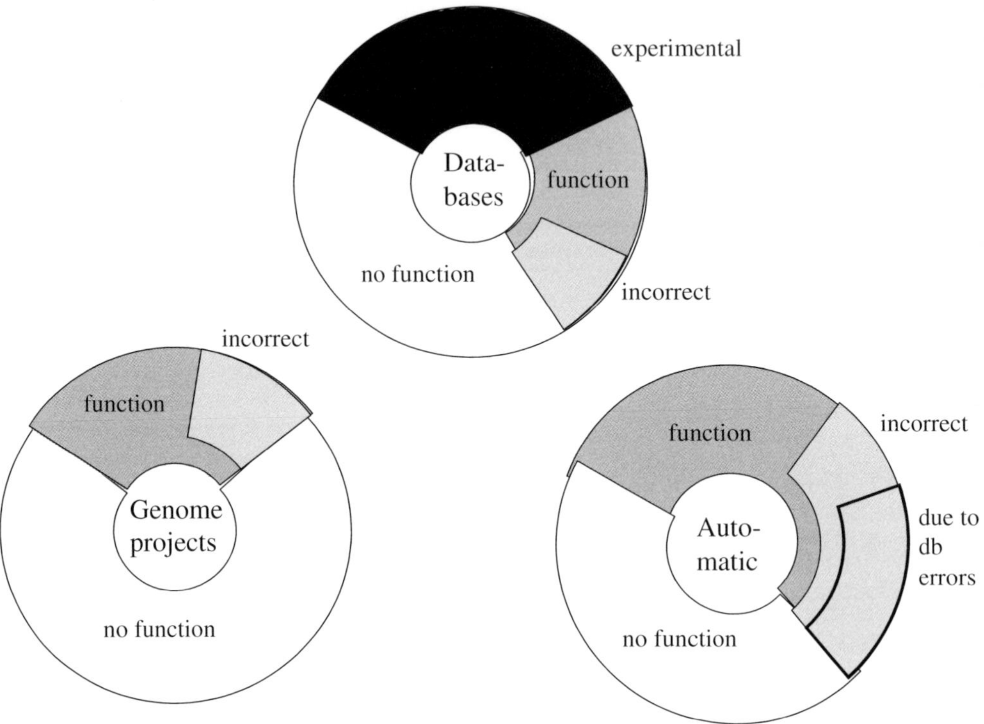

Figure 5. Estimation of the current state of databases and of automatic annotation (via GeneQuiz) regarding the accuracy of function annotation. Top: a 1/4 of the database sequences have some experimental evidence of function, with another 1/4 annotated by similarity to those. However, 1/4 of the latter might be wrongly annotated (half of these wrong annotations originating from genome projects). Left: sequences from genomic projects in databases display a conservative estimation of function for 1/3 of sequences (usually only via similarity), but with a moderate rate of error, about 1/3 of these annotations. This error rate is mostly due to the novelty of the sequences and changes very much depending on the organism considered. Right: automatic annotation from GeneQuiz (as estimated in the example case presented in this chapter) provides more annotations, slightly over 1/2 of the sequences, but with a higher error rate (1/2 of them). However, 2/3 of the errors originate from errors already present in the databases.

of the family followed by phylogenetic analysis, (see above) should be carried out periodically by expert reviewers (see, for example, the MEROPS database of protease families; Barrett *et al.*, 2001). The resulting annotations should be included immediately in the database. Also, the domain organisation of the proteins within a family should be identical. This can be easily detected by checking the length consistency between members of a family, but, especially, using in this comparison the quickly improving protein domain databases, which enable verifying whether source and query sequences have identical domain architecture. Links to these databases are readily introduced in new database entries (Apweiler *et al.*, 2000) making this comparison trivial. With better annotated sequence databases it should become possible to mark the most interesting families in need of experimental characterisation. We are about to enter the era of large-scale biochemical characterisation projects (e.g. Martzen *et al.*, 1999) that (together with structural genomics projects, Burley *et al.*, 1999; Aloy *et al.*, 2001) should optimise the coverage of protein function space on protein sequence space. Before that, we need the most accurate annotation that computation can give.

Acknowledgements

I want to thank the developers of sequence databases, and the current GeneQuiz team at EBI (Cambridge, UK): Paul Janssen, Sophia Tsoka, Ioannis Iliopoulos, Anton Enright, and Christos Ouzounis. I also thank Jaap Heringa, Patrick Aloy, and Carolina Perez-Iratxeta for their helpful comments to the manuscript.

References

Aloy, P., Querol, E., Aviles, F.X., and Sternberg, M.J. 2001. Automated structure-based prediction of functional sites in proteins: applications to assessing the validity of inheriting protein function from homology in genome annotation and to protein docking. J. Mol. Biol. 311: 395-408.

Altschul, S.F., Gish, W., Miller, W., Myers, E.W.,and Lipman, D.J. 1990. Basic local alignment search tool. J. Mol. Biol. 215: 403-410.

Andrade, M.A., Brown,. N.P., Leroy, C., Hoersch, S., de Daruvar, A., Reich, C., Franchini, A., Tamames, J., Valencia, A., Ouzounis, C., and Sander, C. 1999. Automated genome sequence analysis and annotation. Bioinformatics. 15: 391-412.

Apweiler, R., Attwood, T.K., Bairoch, A., *et al.* 2000. InterPro - an integrated documentation resource for protein families, domains and functional sites. Bioinformatics 16: 1145-1150.

Ashburner, M., Ball, C.A., Blake, J.A., *et al.* 2000. Gene ontology: tool for the unification of biology. The Gene Ontology Consortium. Nat. Genet. 25: 25-29.

Bairoch, A, and Apweiler R. 2000. The SWISS-PROT protein sequence database and its supplement TrEMBL in 2000. Nucleic Acids Res. 28: 45-48.

Barrett, A.J., Rawlings, N.D., and O'Brien, E.A. 2001. The MEROPS database as a protease information system. J. Struct. Biol. 134: 95-102.

Bateman, A., Birney, E., Durbin, R., Eddy, S.R., Howe, K.L., and Sonnhammer E.L. 2000. The Pfam protein families database. Nucleic Acids Res., 28: 263-266.

Bork, P., Bairoch, A. 1996. Go hunting in sequence databases but watch out for the traps. Trends Genet. 12: 425-427.

Borthakur, D., and Gao X. 1996. A 150-megadalton plasmid in *Rhizobium etli* strain TAL182 contains genes for nodulation competitiveness on Phaseolus vulgaris L. Can. J. Microbiol. 42: 903-910.

Burley, S.K., Almo, S.C., Bonanno. J.B., Capel, M., Chance, M.R., Gaasterland, T., Lin, D., Sali, A., Studier, F.W., and Swaminathan, S. 1999. Structural genomics: beyond the human genome project. Nat. Genet. 23: 151-157.

Dandekar, T., Huynen, M., Regula, J.T., Zimmermann, C.U., Ueberle, B., Andrade, M.A., Doerks, T., Sánchez-Pulido, L., Snel, B., Suyama, M., Yuan, Y.P., Herrmann, R., and Bork, P. 2000. Re-annotating the *Mycoplasma pneumoniae* genome sequence: adding value, function and reading frames. Nucleic Acids Research 28: 3278-3288.

Devos, D., and Valencia A. 2000. Practical limits of function prediction. Proteins 41: 98-107.

Devos, D., and Valencia A. 2001. Intrinsic errors in genome annotation. Trends Genet. 17: 429-431.

Frishman, D., Albermann, K., Hani, J., Heumann, K., Metanomski, A., Zollner, A., and Mewes, H.W. 2001. Functional and structural genomics using PEDANT. Bioinformatics 17: 44-57.

Gaasterland, T., and Sensen, C.W. 1996. MAGPIE: automated genome interpretation. Trends Genet. 12: 76-78.

Galperin, M.Y., and Koonin, E.V. 1998. Sources of systematic error in functional annotation of genomes: domain rearrangement, non-orthologous gene displacement and operon disruption. In Silico Biol. 1: 55-67.

Hegyi, H., and Gernstein, M. 2001. Annotation transfer for genomics: measuring functional divergence in multi-domain proteins. Genome Res. 11: 1632-1640.

Hoersch, S., Leroy, C., Brown, N.P., Andrade, M.A., and Sander, C. 2000. The GeneQuiz Web server: protein functional analysis through the Web. Trends Biochem. Sci. 25: 33-35.

Iliopoulos, I., Tsoka, S., Andrade, M.A., Janssen, P., Audit, B., Tramontano, A., Valencia, A., Leroy, C., Sander, C., and Ouzounis, C.A. 2000. Genome sequences and great expectations. Genome Biology 2: interactions0001.1-0001.3

Koonin EV, Mushegian AR, Bork P. 1996. Non-orthologous gene displacement. Trends Genet. 12: 334-336.

Martzen, M.R., McCraith, S.M., Spinelli, S.L., Torres, F.M., Fields, S., Grayhack, E.J., and Phizicky, E.M. 1999. A biochemical genomics approach for identifying genes by the activity of their products. Science 286: 1153-1155.

Pearson., W.R. 1996. Effective protein sequence comparison. Methods in Enzymology, 266: 227-258.

Ponting, C.P., Schultz, J., Copley, R.R., Andrade, M.A., and Bork, P. 2000. Evolution of domain families. Advances in Protein Chemistry 54: 185-244.

Scharf, M., Schneider, R., Casari, G., Bork, P., Valencia, A., Ouzounis, C., and Sander, C. 1994. GeneQuiz: a workbench for sequence analysis. Intelligent Systems for Molecular Biology 2: 348-353

Schultz , J., Copley, R.R., Doerks, T., Ponting, C.P., and Bork, P. 2000. SMART: a web-based tool for the study of genetically mobile domains. Nucleic Acids Res. 28: 231-234

Snel, B., Lehmann, G., Bork, P., and Huynen, M. A. 2000. STRING: A web-server to retrieve and display the conserved neighbourhood of genes. Nucleic Acids Res. 28: 3442-3444.

From: *Bioinformatics and Genomes: Current Perspectives*
Edited by: Miguel A. Andrade

Chapter 7

Dynamics and Complexity in Systems Biology Modelling: Theoretical Challenges in Metabolic Simulation

Eric Minch

Abstract

The networks of metabolic and control interactions in the cell challenge our ability to gather appropriate data and to develop appropriate tools, but primarily the challenge is a theoretical one. The system we are trying to understand is not a machine with fixed structure. The components themselves as well as the types of components, and the interaction events in which they participate, are continually changing and altering the control and spatial structure, due to environmental perturbations and autonomous dynamics. This chapter briefly describes the available modelling methodologies (including stochastic, semiquantitative, and hybrid models), and explains the essential conceptual challenges such as:

- What is an appropriate representation of events involving interactions among internal regions of a macromolecule?
- How can we detect and characterize signalling and regulatory activities, and their effect on the control structure?
- At what point must a component be regarded as macroscopic, affecting the spatial structure?
- What is the proper representation of function as opposed to process?

Introduction

Metabolic modelling has never been more promising, more exciting, and more powerful. We have more tools, techniques, and algorithms at our disposal, and the well-known explosion of data from the advances in genomics have provided more data than ever before. We have detailed and fairly accurate representations of various metabolic pathways in different organisms (Kanehisa *et al.*, 2002; Selkov *et al.*, 1998), and holistic approaches which encompass entire organisms in less detail (Karp *et al.*, 1997; Tomita *et al.*, 1999). Metabolic reconstruction promises eventually to deliver models which are both detailed and integrative (Bork *et al.*, 1998), while metabolic engineering promises to deliver methods for designing modifications to organisms tailored to the production of inexpensive, efficient agricultural and pharmaceutical products (Stephanopoulos *et al.*, 1998). As the scope and power of these methods expand, so do the horizons of our expectations. But this is still mostly promise, and many difficulties lie ahead. Although genomics has produced a flood of information on genes, proteins, enzymes, and the regulatory and signalling mechanisms to control development, differentiation, and adaptation to environmental fluctuations, the problem of identifying relevant facts requires sophisticated methods of data mining and knowledge base construction. We will not discuss these here, as many efforts are under way to solve this problem, and we can reasonably hope for progress in its solution.

Furthermore, detailed and complete modelling requires detailed and complete data on kinetics, and these data are still quite sparse. Filling these gaps can entail labor intensive processes requiring weeks of effort by laboratory scientists, or compute-intensive processes requiring extensive calculation using sophisticated algorithms to model folding, docking, and detailed conformation and charge distribution etc. Given sufficient time and resources, these answers too can be obtained, so we will not discuss these either. The focus in this chapter is on techniques for metabolic modelling based on datasets with appropriate relevance, accuracy, and completeness.

Dynamics And Function

The overall goal of metabolic modelling is to reflect as realistically as possible the underlying physico-chemical events of metabolism, while maintaining contact with the functional biological observables familiar to investigators. Events relevant to metabolism can range from – at the low end – conformational changes taking a fraction of a nanosecond to – at the high end – intercellular signalling mechanisms or cell cycle regulatory processes occurring over a period of many hours, encompassing a dynamic range of at least fourteen orders of magnitude. Including yet finer detail of reaction mechanisms or broadening the scope to developmental (or even evolutionary) processes increases this range.

At one extreme, then, an *ab initio* based approach is conceivable, in which the configuration of each molecule and its interactions with other molecules is calculated in detail from fundamental principles. The opposite extreme is a knowledge-based textbook level representation, in which the cause and effect relations among the various high-level objects and transformation events are stored as rules, and the system progresses from one state to the next by the logical implications of these rules. These two extreme positions define respectively the quantitative and the qualitative approaches in simulation methodology. In practice, we are usually interested only in a limited range of events. The more gradual the changes over this range, the more appropriate it is to model them quantitatively and microscopically; events which are discrete or infrequent are more suitably treated with qualitative, macroscopic methods.

Metabolic modelling is only rarely concerned with detailed mechanisms of individual reactions, so it is possible to deal with pools of metabolites instead of individual molecules. The basic event then is the reaction, but here "reaction" is defined not as an interaction between individual molecules but rather as the conversion of some quantity of one pool into a stoichiometrically determined quantity in another pool. The simplest models assume a single well-mixed compartment, and can ignore spatial considerations (i.e., both the pools and the compartment in which they take place can be considered as dimensionless points). Multicompartment models augment this schema with intercompartment diffusion and/or transport events. Both quantitative and qualitative approaches have their valid uses. The main factors determining the approach are the sort of questions that are to be asked and the data available (the type, the amount, and the quality). Difficulties arise when simulating two or more levels of objects and events. So long as the qualitative structure of the system can be assumed constant, it is possible, though by no means trivial, to allow translation back and forth between different levels of description. One can, for example, model enzymatic catalysis of a reaction using either a detailed, linear, kinetic description of the various substrates and enzymatic complexes, or, alternatively, a nonlinear Michaelis-Menten (or other mechanism) description, or even a logic-based switching description.

More generally, though, we are challenged by systems in which the qualitative structure can change over time as a result of changing conditions, whether in response to toxins, to therapeutic interventions, or to processes of cellular development and differentiation. For these cases, there is as yet no general method which allows both identification of macroscopic objects and quantitative elucidation of underlying events. To address this need, two additional types of simulation, augmenting the purely quantitative and the purely qualitative, have arisen. The semi-quantitative approach is based on a qualitative model, but allows incorporation of available quantitative data in the model state; continuous processes can then be switched by discrete events, and these in turn can be modulated, via thresholding functions, by the quantitative state. Hybrid approaches have also been proposed, which weld together quantitative and qualitative descriptions of the same phenomenon, albeit without allowing translation between the two. In the next section, we briefly describe and compare examples of these four approaches.

Approaches To Metabolic Modelling

Quantitative Approach

Continuous kinetic models were the first to be applied to limited systems of chemical reactions, as there was already an extensive history of dynamical systems simulation using differential equations. Use of this paradigm has continued, especially with increasing computational power over the last decades (Ziegler and Weinberg 1970), and some of the most recent efforts are based on pure quantitative models.

Several computer packages have been specifically developed for quantitative metabolic modelling (i.e., for kinetic simulation and optimization), including MIST (Ehlde and Zacchi, 1995), DBSolve (Goryanin *et al.*, 1999), SCAMP/Jarnac (Sauro, 1993), and Gepasi/COPASI (Mendes and Kell, 1998). Without intending a comparative review of these, we can take Gepasi as an example: it is well-maintained, quite flexible, and easy to use, has been used for dozens of publications, and exemplifies the state of the art for quantitative modeling. It allows specification of a variety of reaction mechanisms, can search for solutions and estimate parameters based on empirical data, and characterize steady states using metabolic control analysis

(MCA) and linear kinetic stability analysis. Running a simulation using the program involves specifying the compartments (if more than one) and chemical reactions, marking which substances are buffered, choosing or defining a kinetic type for each reaction (e.g., reversible or irreversible, Michaelis-Menten or Hill, Uni-Uni or Bi-Bi, etc.), and setting the initial conditions. A number of other controls allow varying the time step, the integration technique, the variables to be displayed and/or logged, the optimization method (when appropriate), and so on. Only four variables can be displayed during running, but any variable or parameter of the model, including user-defined transformations, can be sent to an output file for later analysis.

An even more recent example applying this paradigm is E-CELL (Tomita *et al.*, 1999), in development since 1996 at Keio University by an interdisciplinary team of about thirty scientists and software engineers. E-CELL is, like Gepasi, a general-purpose biochemical simulation environment, which has been used to model the primary metabolism of a whole cell (an *in silico* chimera of *M. genitalium* and *E. coli*), resulting in a counterintuitive prediction of the glucose starvation response which still awaits empirical verification. Although the model does not include replicative or regulatory mechanisms, the E-CELL framework is claimed to be eventually capable of including these as well by employing dynamically-varying local time steps for nonlinear reactions.

Another approach within the general category of continuous simulation is stochastic modelling. The above examples permit only deterministic processes, but in modelling signalling and regulatory pathways the relevant molecular species are often found only at nanomolar concentrations or less, i.e., down to a small number of copies per cell. Whether or not these react, and at exactly what time point they react, can make a qualitative difference to the cell's fate. Although the processes modeled are discrete, stochastic simulation can be regarded as a continuous method because it is based on continuous probability distributions.

The most widely used approach to stochastic simulation is one or another variation on the Gillespie algorithm (Gillespie, 1976), in which reactions are selected randomly under constraint of their rates and the concentration of their reactants. Gibson's (Gibson and Bruck, 2000) more recent and highly efficient "Next Reaction" method has been applied to regulation of *E. coli* genes by the bacteriophage λ lysis/lysogeny decision circuit (Gibson and Bruck, 2001) (see Section "Hybrid approaches" below). These methods successfully handle processes in which low concentrations can still have important effects, but to do so they must be run as Monte Carlo experiments with a large number of replicates. An alternative method for stochastic simulation, which has seen little use in metabolic modeling, delivers analytical solutions based on stochastic differential equations.

Whether by numerical or analytical solution of equations or by Monte-Carlo iteration, simulating these systems is very compute-intensive; when it is necessary to model at different time scales, a step size suitable for the shortest time scale is far too demanding for longer time scales. Furthermore, numerical integration is highly sensitive to nonlinearities, thus ill-suited to modelling both kinetic and regulatory or signalling interactions together. Finally, the necessary kinetic data are still sparse, so the majority of parameters must be estimated or default values.

Qualitative Approach

An intimation of the purely qualitative approach can be seen in the work of Sugita (Sugita, 1963), who pointed out the essential switching logic nature of enzymatically catalysed reactions*. Use of knowledge bases and production systems for biochemical simulation was pioneered by Karp (Karp and Riley, 1993), and finds its fullest expression in the EcoCyc system (Karp *et al.*, 1997). A detailed summary of some of this work is also included in the review by Galper and Brutlag (1994).

Using the data of EcoCyc, Heidtke and Schulze-Kremer have developed a qualitative simulation framework (BioSim) (Heidtke and Schulze-Kremer, 1998) and a model description language (MDL) (Heidtke and Schulze-Kremer, 1999), to derive qualitative models from the EcoCyc knowledge base. The Lisp-based frame definitions within EcoCyc were translated to Prolog, then the classes were defined within MDL as being either objects (compounds, elements, enzymes, genes, proteins) or processes (reactions, pathways). Thus a model could be constructed on the fly from a query to the EcoCyc knowledge base. An interpreter was then applied to the model to generate qualitative behavior (i.e., indications of the increase or decrease in quantity of a substance). In an example, the knowledge base is queried for phosphoglucose isomerase (EC 5.3.1.9), the enzyme is retrieved, and its data are used to generate a model consisting of three objects (the enzyme itself plus its two substrates) and four processes (formation and breakdown of the enzyme-substrate complex, an overall reaction combining these, and an additional process describing the reaction in absence of the enzyme); the resulting behavior indicates the qualitative changes (i.e., directions without

* The switching logic nature of genetic regulation and signal transduction cascades was still new at the time: research leading to the discovery of the operon continued up until the early 1960s, and the Nobel prize for this work was not awarded until 1965.

rates or durations) in concentrations of reactant, product, complex, and free enzyme over a period of six arbitrary time steps from initial state to final equilibrium.

It should also be mentioned in this context that a good many efforts are under way in reconstructing and representing genetic regulatory networks (de Jong *et al.*, 2002). Although these do not model metabolism *per se*, they do constitute qualitative models whose function will eventually be necessary for accurate metabolic models.

The drawback of the qualitative approach is that the quantitative aspects of the system have been abstracted away, so that accurate modelling of gradual change is impossible; even to approximate it is difficult without considerable complication of the knowledge- and rule-base. Variations in concentration of metabolites, in rate of reactions, or in delay between events, cannot be both easily and accurately modeled using purely qualitative methods.

Semi-Quantitative Approach

One influential model used in the semi-quantitative approach is the Petri net formalism, first employed for metabolic modelling by Reddy in a purely qualitative form (Reddy *et al.*, 1993). A Petri net is a bipartite graph i.e., a graph in which the vertices are partitioned into two subsets, such that vertices from each set are connected only to vertices in the other set. For purposes of metabolic modeling, one set of vertices (called places) is used to represent metabolites; the other set of vertices (called transitions) is used to represent reactions. The characteristic of Petri nets which is useful for modelling is that the places contain "tokens", and when every place incident to a transition is occupied by a token, the transition fires, which changes the state of each of the places connected to the transition.

Petri nets have been extended in many different ways (black/white vs. colored, delayed or timed, capacity, self-modifying, stochastic, etc.). By incorporating one or more of these extensions, it is possible to model at any level or combination of levels from qualitative to quantitative, i.e., one can include as much quantitative information as available on concentrations, rates, etc. Two of the most recent efforts are those of Hofestaedt and Thelen (1998) and of Matsuno *et al.* (2000).

The extensions required by the Hofestaedt and Thelen model were

- allowing multiple tokens at each place (corresponding to the concentrations of metabolites)
- modifying the firing rule of the transitions (reflecting the stoichiometry of the reactions)

- limiting the capacity of the places (which is not actually needed for modelling but only for obtaining analytical results on the behavior of the Petri net)
- making the arrow weights or delays depend on the place token counts (so that reaction rates depend on concentrations)
- replacing simple arrow weights with functions (to allow complex kinetic mechanisms)

Using these extensions, they found it possible to model both qualitative and quantitative processes at several time scales, including activation, inhibition, and transcription regulation, and demonstrated these in an example model of glycolysis.

Matsuno *et al.* (2000) extended the places and transitions to allow continuous as well as discrete values, allowing a model of the phage λ lysis-lysogeny circuit (see Section "Hybrid Approaches" below) that includes analog, digital, and potentially also stochastic processes to be represented in a single formalism.

The other main contender in the realm of semi-quantitative modelling is the QSIM/Q2/Q3/NSIM family of modelling languages. These encompass a series of developments stemming from Kuipers' first LISP-based implementation of a qualitative simulation engine (Kuipers, 1994). They are characterized by relying on landmarks and intervals: landmarks are regions of the state-space in which qualitative changes in direction can occur (e.g., separatrices between attractors), while intervals bound the set of possible solutions, both in time and in the values of variables. In the limit, i.e. as the intervals approach zero, Q3 (Berleant and Kuipers, 1997) and NSIM (Kay and Kuipers, 1993) are equivalent to purely quantitative ODE-based modeling. In other words, they offer all the power of quantitative models but degrade gracefully under uncertainty in parameter values.

While flexible, the formal frameworks of the semi-quantitative approach are still unfamiliar, and not yet very well standardized. In the case of Petri nets, the more accurately quantitative the model is to be, the more extensions must be made to the supporting formalism, making it somewhat complex and unwieldy. Nonetheless, the representational power and generality of these techniques, particularly the landmark/interval paradigm, means that they will probably be used more often in the future.

Hybrid Approaches

All of the above models are integral, in the sense that each consists of a single set of objects and of relations among these objects. A differential equation representation, for example, consists of a set of variables and a set of differentials defined on the variables. Production systems and switching networks are relatively simple formal machines, i.e., boolean variables linked by logic gates; the Petri net is a yet more complex class of machine, even with all of its various extensions: one can add delays, introduce asynchronous lines or stochastic processes, but it is still a single model. A hybrid model, though, while still formal, contains two or more classes of objects, such that the interactions among the members of each class are defined according to a set of rules peculiar to that class (constituting an integral model), while the interactions between classes (the "glue" linking the various integral representations) are specified by *ad hoc* rules peculiar to the model (Alur *et al.*, 2002). Hybrid models are thus non-generic, special-purpose models which must be hand-crafted for each target system.

McAdams and Shapiro (1995) present an outstanding example of this approach, in which they simulate the kinetics and signal logic of the bacteriophage λ lysis/lysogeny decision circuit. Their model involves twelve genes organized in five operons with seven promoters, and both mRNA, direct gene products, and posttranscriptional modifications. The kinetics of transcription and signaling are modeled using numerical integration of differential equations, while the switching logic is superimposed on the signal levels using both sigmoid-thresholded and edge-triggered boolean gates. Signal delays are produced by distance between genes on the DNA, transcription rates, concentration thresholds for signalling proteins, and rates of protein degradation. The model is remarkably detailed and accurate, and allows verification of the logic as well as identification of the control function of different components.

Another form of hybrid simulation method is implicit in metabolic control analysis and related techniques (Schilling *et al.*, 1999). These techniques *per se* do not directly realize simulations. Instead, they derive analytical solutions from algebraic system descriptions in order to detect discrete characteristics such as stable modes, sets of co-regulated reactions, and bottlenecks, as well as global system parameters such as control coefficients (the extent to which a given enzyme controls a pathway), and elasticity coefficients (the extent to which the activity of an enzyme is modulated by other substances). A computer implementation of these techniques which allowed control of initial conditions and display of state, however, would constitute a hybrid simulation if the discrete solution modes were coupled to discrete or continuous reaction, transport, regulatory, and signalling mechanisms which fed back to create a new set of initial conditions.

One advantage of hybrid models such as this is that (like semi-quantitative models) they can represent aspects of continuous kinetics at short time scales and discrete switching mechanisms at longer time scales. A further advantage, distinguishing them from the semi-quantitative approach, is that they utilize algorithms for these different time scales which have been highly optimized over decades of use, which are both efficient and familiar. The practical disadvantage is that they must be recreated by hand for each particular target system. The theoretical disadvantage is that there is no well-defined mapping between the objects or events at one level and those at another; these must all be defined and inferred *ad hoc* for each case. Developing algorithms to provide this translation automatically would remove both disadvantages.

Let's walk through some of the difficulties faced by someone given the challenge of using current simulation technology and anticipated data to simulate a whole cell realistically, as proposed above in the Section "Dynamics and function".

THEORETICAL CHALLENGES

The most immediate current obstacles to metabolic modelling are the practical problems of obtaining relevant and accurate data (as mentioned in the Introduction). Let's imagine for a moment that all relevant factors have been identified and all necessary parameters have been determined (this situation will inevitably be approached, given sufficient time and resources). Using the available tools described above, can we then deliver to biologists the capabilities they need in order to understand the processes of living cells? The answer can only be a qualified "No". Certainly, to the extent that the control pathways of interest have been curated by a domain expert, we can simulate the signalling and regulatory interactions within – as well as upstream and downstream of – a given metabolic pathway. But the wealth of information implicit in our assumption of complete information will assure that many control pathways – and interactions among control pathways – will exist which we cannot adequately detect and characterize with our current toolset.

The reason – and it is a very deep reason – is that we are trying to model the modelling relation itself. All control processes involve a semiotic component: measurement, communication, and control involve symbolic interactions reflecting the current state of the cell and its surroundings. A symbolic interaction is more appropriately modeled with a state transition table than with a set of ODEs. The states in the state transition table are not algorithmically derivable from the underlying physical state set. If we attempt to represent these states and tables at a physical level, i.e., in terms of the detailed physiochemical interactions underlying them, we lose contact with the functional symbolic interpretation. In other words, if our completely-specified system is simulated, then all of the control interactions will appear to be no more than mass-transfer interactions, unless we either curate them into control status or somehow automatically determine their control status. Once we know the signalling and regulatory pathways we can model them quantitatively (with ODEs or hybrid or stochastic models). But there will undoubtedly be many control pathways we are still unaware of hidden in these interactions*. How do we detect and characterize control pathways as distinct from metabolic pathways, if our model involves only mass transfer interactions? Alternatively (the inverse problem), how can we represent the continuous interactions underlying a functional switching mechanism? These two problems together are known as the problem of the "epistemic cut" (Pattee, 1997).

Transport

Although most ODE simulations are framed in terms of reactions, transport processes do not fit simply in such a framework. The proper framework to include such processes is the reaction-diffusion system, in which the reactions have their characteristic kinetic mechanisms and parameters (mass action, Michaelis-Menten, Hill, Bi-Uni, Ping-Pong, etc.), and the diffusion processes have their mechanisms and parameters (passive diffusion, port, export, antiport, symport, etc.). Transport thus poses no severe theoretical problem, but requires that the model be expanded to include multiple compartments, each with its own reaction network, and that the reactions be somehow tagged by the compartment in which they take place. This is simple in concept, but there is still no standard classification of such compartments, nor of the method of representing intercompartmental processes (although the Gene Ontology (The Gene Ontology Consortium, 2001) and the Systems Biology Markup Language (Hucka *et al.*, 2002) indicate progress).

Polymerization

Representing polymerization processes brings with it a number of difficulties which appear in several other problem areas (see Sections " Conformations, complexes, domains" and "Genes" below). In the most simple cases, e.g., polylactate synthesis, there is a single unit of polymerization and a single mechanism for the lengthening of the polymer. In more complex cases, we

* The human genome is currently estimated to contain only about 30K-40K ORFs, about double the number of *D. melanogaster* or *C. elegans*; most of the difference is presumed to lie in the proteins responsible for signalling events in the cellular environment and regulating gene transcription. The number of control pathways is estimated to be orders of magnitude greater than the number of the metabolic pathways (Csete and Doyle, 2002).

may have multiple units of polymerization (as in DNA, RNA, and proteins), or multiple mechanisms for the extension of the polymer (as in glycogen synthesis (Melendez *et al.*, 1999)).

This is where the conflict begins between the strategy of modelling the elements of the system as unique entities (in which case each size of the polymer would be explicitly represented), and that of a state representation (in which case the polymer might be present in different size-states each with its own probability). In the state-based solution, the concentration of the polymer can be represented as a probability distribution over the sizes and branchings, but this requires some modification of a straightforward ODE-based simulation, involving stochastic differential equations or Monte Carlo evaluation.

Largely, though, this obstacle seems to be a matter of computational resources: with enough memory and enough CPU cycles either the state-based or the species-based models would be feasible; there appears to be no overriding theoretical justification for preferring one over the other.

Conformations, Complexes, Domains

Conformational changes pose a dual puzzle. On the one hand, we have the problem of representing these as events within the framework of a reaction model, since there may be a change in state without any reaction. On the other hand, we again have the problem of representing a probability distribution over states versus the deterministic production of explicit species. In the chain of events between recognition at receptors of extracellular events and the eventual regulation of gene transcription in the nucleus, many different complexes (protein-ligand, protein-protein, protein-DNA) may be formed. For example, EGF (epidermal growth factor) binds with the extracellular domain of the EGF receptor EGFR; a pair of such heterodimers complexes within the cytoplasmic membrane; the inner tails of the complexed EGFR contains a kinase domain, and one of these binds with free cytoplasmic ATP; this complex then participates in mutual phosphorylation with a neighboring EGFR complex; finally, the phosphorylated complex binds with cytoplasmic complexes to initiate a signal cascade terminating eventually in the nucleus with gene regulatory effects. Denoting each complex with its own unique identifier and database entry seems impractical, yet each must somehow be distinguished from previous and successor species. In addition, the combination of many domains subject to phosphorylation or other modifications results in a potentially very large number of distinct states to be considered.

Although it is undoubtedly conceptually simpler and computationally more efficient to represent these modifications as states rather than as distinct chemical species, this poses a theoretical problem, since the states and their characteristics will always have to be explicitly listed; any interaction not conforming to the listed states will be unavailable. This problem does not exist for the approach which represents each species explicitly, since the set of distinct species may be completely open-ended. This is the basic problem of the epistemic cut, with the added twist that we wish to allow the user to apply the cut at any desired level. Resolving the conflict between the practical necessity of an efficient representation and the theoretical desire for an open-ended set of "states" is a challenge with no commonly agreed-upon answer.

Genes

The problem here is similar to that of complexing and conformational change, in that a region of the DNA molecule is involved in interactions (promoter recognition, polymerase attachment, initiation and elongation of transcript). Just as we can consider conformational changes as involving interactions among domains within a protein molecule, genes (transcription units) and modifiers (promoters, enhancers) can be considered as interacting individually with ligands or proteins, etc. A plausible approach is to regard each as a chemical species *per se*, but this ignores the neighborhood relations among genes, distorting the reality of the interactions themselves, since agents binding to one site of the chromosome undoubtedly affect not only that site but upstream and downstream regions as well.

Signalling and Regulation

Here is the major challenge. Signalling and regulatory interactions involve control processes. Successful modelling of control systems entails being able to represent a control process as both a digital switching network – to emphasize its nature as a control process – and as a continuous analog system – to investigate the dynamical and quantitative characteristics of the process. Does the dual nature of control processes entail a corresponding duality in their representation?

How do we bridge the gap between mass-transfer physiochemical interactions and symbolic control processes? As we have seen above in Section "Approaches", both the semiquantitative and the hybrid methods can model the dual nature of control processes, delivering both views. The hybrid method, though, will not allow translating back and forth between views, i.e., the analog component models the continuous metabolic processes and the digital component models the discrete control processes. The semiquantitative method allows such translation, so that we can look inside a control process to examine the kinetic mechanisms, if this information is available. Currently there is only one way to provide it, namely by experiments and human curation of the results into hypothetical mechanisms. Curation works, but it is slow, it is labor-intensive, and it will inevitably miss important cause-effect interactions. The alternative is an algorithmic approach: ideally this would propose hypothetical mechanisms for observed net effects, and would also infer high-level control processes from known kinetic, stoichiometric, and topological relations. Such a translation would even permit on-the-fly generation of hybrid models. We do have some rules and heuristics for detecting control processes (Pattee, 1997; Minch, 1998; 2000), for example:

- signs are thermodynamically degenerate
- changes in signal state are separated by low energy barriers
- signs and records must be stable within their operating cycle
- metabolic cost (in matter and energy) of control processes is low relative to cost of the process being controlled

These theory-based rules have not yet been widely applied and thus have not demonstrated their practical utility; there is much scope here for future efforts.

Evolving Space

This is basically unsolved. The problem is that some of the processes to be modeled include constraints imposed e.g. by spatial structure, and that some processes themselves result in changes in spatial structure. The large functional complexes (e.g. ribosomes) are an example, as are processes such as development of cellular ultrastructure. This could be generalized to nonspatial constraints, such that all cases of mutual causality (downward constraints and upward emergence) were included, thus encompassing the duality problem. No system for cellular simulation that I am aware of even addresses this problem, much less provides a solution, although work on modelling tissue differentiation (Shapiro and Mjolsness, 2001), colony growth (Kreft *et al.*, 2000), or tumor formation (Kansal *et al.*, 2000) using cellular automata with variable or nonuniform grids might be extended for this purpose.

Conclusion

The practical problems of obtaining relevant and accurate data will eventually yield to sustained effort of dedicated resources. Some of the easier problems of representation can be alleviated by current or future powerful computational capabilities. The harder problems of representing polymerization processes, and of reconciling the state-based vs. object-based representations for them, can be resolved by convention: there appear to be no theoretical difficulties. Resolving this question for complexes or for control processes, however, means deciding where to apply the epistemic cut. This can also be decided by convention or convenience within any problem domain, but it will typically vary between problem domains, and there appears no easy way to allow this. Some of the work towards algorithmic definition of functional subsystems (e.g., Schilling *et al.*, 1999; Ettema *et al.*, 2001) may help in providing such decisions automatically. The new field of Biosemiotics (Hoffmeyer, 1996) holds out promise, and may provide a principled approach to deciding where to divide the world between physics and control. We want to provide these decisions automatically, not only to spare domain experts the effort of recording these characterizations, but also to give them freedom to apply them at whatever level is most appropriate to the data and the problem: to translate freely back and forth between the symbolic and the dynamic interpretation of the same phenomenon.

Resolution of these theoretical problems cannot come from computer scientists or mathematicians, nor from philosophers, nor yet from chemists, molecular biologists, physiologists, or geneticists; they will be resolved only by people with all of these perspectives working together.

References

Alur, R., Belta, C., Kumar, V., Mintz, M., Pappas, G.J., Rubin, H., and Schug, J. 2002. Modelling and analyzing biomolecular networks. IEEE Computing in Science and Engineering 4: 20-31.

Berleant, D., and Kuipers, B. 1997. Qualitative and quantitative simulation: Bridging the gap. Artificial Intelligence 95: 215-256.

Bork, P., Dandekar, T., Diaz-Lazcoz, Y., Eisenhaber, F., Huynen, M., and Yuan, Y. 1998. Predicting function: from genes to genomes and back. J. Mol. Biol. 283: 707-725.

Csete, E.M., and Doyle, J. 2002. Reverse engineering of biological complexity. Science 295: 1664-1669.

de Jong, H. 2002. Modelling and simulation of genetic regulatory systems: a literature review. J. Comput. Biology. 9(1): 67-104.

Ehlde, M., and Zacchi, G. 1995. MIST: a user-friendly metabolic simulator. Comput. Appl. Biosci, 11: 201-207.

Ettema, T., van der Oost, J., and Huynen, M. 2001. Modularity in the gain and loss of genes: applications for function prediction. Trends in Genetics 17: 485-487.

Galper, A.R., and Brutlag, D.L. 1994. Computational simulations of biological systems. In: Biocomputing: Informatics and Genome Projects. Academic Press, San Diego. p. 269-305.

Gibson, M.A., and Bruck, J. 2000. Efficient exact stochastic simulation of chemical systems with many species and many channels. J. Phys. Chem. A. 104: 1876 -1889.

Gibson, M.A., and Bruck, J. 2001. A probabilistic model of a prokaryotic gene and its regulation. In: Computational Modelling of Genetic an Biochemical Networks. J. Bower and H. Bolouri, eds. MIT Press, Cambridge (MA). p. 49-73.

Gillespie, D.T. 1976. A general method for numerically simulating the stochastic time evolution of coupled chemical reactions. J. Comput. Phy. 22: 403-434.

Goryanin, I., Hodgman, C., and Selkov, E. 1999. Mathematical simulation and analysis of cellular metabolism and regulation. Bioinformatics 9: 749-758.

Heidtke, K.R., and Schulze-Kremer, S. 1998. BioSim – a new qualitative simulation environment for molecular biology. In: Proceedings of Sixth International Conference on Intelligent Systems for Molecular Biology. AAAI Press, Menlo Park (CA). p. 85-94.

Heidtke, K.R., and Schulze-Kremer, S. 1999. Deriving simulation models from a molecular biology knowledge base. In: Proceedings of Fourth Workshop on Engineering Problems for Qualitative Reasoning. Sixteenth International Joint Conference on Artificial Intelligence. Stockholm, Sweden. July 31-August 6.

Hofestaedt, R., and Thelen, S. 1998. Quantitative modelling of biochemical networks. In Silico Biology, 1: 39-53.

Hoffmeyer, J. 1996. Signs of Meaning in the Universe. University Press, Bloomington, Indiana.

Hucka, M., Finney, A., Sauro, H.M., Bolouri, H., Doyle, J.C., Kitano, H., Arkin, A.P. *et al.* 2002. The System Biology Markup Language (SBML): a medium for representation and exchange of biochemical network models. Bioinformatics. In Press. See also http: //www.cds.caltech.edu/ erato/sbml/docs/

Kanehisa, M., Goto, S., Kawashima, S., and Nakaya, A. 2002. The KEGG databases at GenomeNet. Nucleic Acids Research 30: 42-46. See also http: //www.genome.ad.jp/kegg/

Kansal, A.R., Torquato, S., Harsh, G.R., Chiocca, E.A., and Deisboeck, T.S. 2000. Cellular automaton of idealized brain tumor growth dynamics. BioSystems 55: 119-127.

Karp, P.D., and Riley, M. 1993. Representations of metabolic knowledge. In: Proceedings of First International Conference on Intelligent Systems for Molecular Biology. AAAI Press, Menlo Park (CA). p. 207-215.

Karp, P., Riley, M., Paley, S., Pellegrini-Toole, A., and Krummenacker, M. 1997. EcoCyc: electronic encyclopedia of *E. coli* genes and metabolism. Nucleic Acids Research 25: 43-50; See also http://BioCyc.org/ecoli/

Kay, H., and Kuipers, B. 1993. Numerical behavior envelopes for qualitative models. In: Proceedings of the 11th National Conference on Artificial Intelligence. AAAI Press/MIT Press, Menlo Park (CA)/Cambridge (MA). AAAI-93: p. 606-613.

Kreft, J.-U., Booth, G., and Wimpenny, J.W.T. 2000. Applications of individual-based modelling in microbial ecology. In: Microbial Biosystems: New Frontiers. Proceedings of the 8th International Symposium on Microbial Ecology. August 9-14, 1998, Halifax, Canada. Bell, C.R., Brylinsky, M., and Johnson-Green, M., eds. Atlantic Canada Soc. Microbial Ecology, Halifax, Canada. p. 917-923.

Kuipers, B. 1994. Qualitative Reasoning: Modelling and Simulation with Incomplete Knowledge. MIT Press, Cambridge (MA). See also http: //www.cs.utexas.edu/users/qr/QR-software.html/

Matsuno, H., Doi, A., Nagasaki, M., and Miyano, S. 2000. Hybrid Petri net representation of gene regulatory network. Pacific Symposium on Biocomputing. 5: 338-349.

McAdams, H.H., and Shapiro, L. 1995. Circuit simulation of genetic networks. Science 269: 650-656.

Meléndez, R., Meléndez-Hevia, E., and Canela, E.I. 1999. The fractal structure of glycogen: a clever solution to optimize cell metabolism. Biophys. J. 77: 1327-1332.

Mendes, P., and Kell, D.B. 1998. Non-linear optimization of biochemical pathways: applications to metabolic engineering and parameter estimation. Bioinformatics 14: 869-883. See also http: //www.gepasi.org/

Minch, E. 1998. The beginning of the end: on the origin of final cause. In: Evolutionary Systems. G. Van de Vijver, ed. Kluwer Academic Publishers, Dordrecht.p. 45-58.

Minch, E. 2000. The Rosetta stone of stability. Annals of the New York Academy of Sciences 901: 148-154.

Pattee, H.H. 1997. The physics of symbols and the evolution of semiotic control. Workshop on Control Mechanisms for Complex Systems: Issues of Measurement and Semiotic Analysis. Las Cruces, New Mexico. Dec. 8-12,1996. Santa Fe Institute Studies in the Sciences of Complexity, Proceedings Volume. Addison-Wesley, Redwood City (CA).

Reddy, V.N., Mavrovouniotis, M., and Liebman, M. 1993. Petri net representation in metabolic pathways. In: Proceedings of First International Conference on Intelligent Systems for Molecular Biology. AAAI Press, Menlo Park (CA). p. 328-336

Sauro, H.M. 1993. SCAMP: a general-purpose simulator and metabolic control analysis program Comp. Appl. Biosci. 9: 441-450. See also http: //www.cds.caltech.edu/~hsauro/

Schilling, C.H., Schuster, S., Palsson, B.O., and Heinrich, R. 1999. Metabolic pathway analysis: basic concepts and scientific applications in the post-genomic era. Biotechnology Progress 15: 296-303.

Selkov, E., Jr., Grechkin, Y., Mikhailova, N., and Selkov, E. 1998. MPW: the metabolic pathways database. Nucleic Acids Research 26: 43-45. See also http: //wit.mcs.anl.gov/WIT2/

Shapiro, B.E., and Mjolsness, E.D. 2001. Developmental simulations with Cellerator. Proceedings of The Second International Conference on Systems Biology, November 4-7, 2001. Yi, T.-M., Hucka, M., Morohashi, M., and Kitano, H., eds. p. 342-351.

Stephanopoulos, G., Aristodou, A., and Nielsen, J. 1998. Metabolic Engineering: Principles and Methodologies. Academic Press, San Diego. See also http: //www.ibt.dtu.dk/cpb/metaboleng/metabol.htm

Sugita, M. 1963. Functional analysis of chemical systems in vivo using a logical circuit equivalent. II: The idea of a molecular automaton. J. Theoretical Biology 4: 179-192.

The Gene Ontology Consortium. 2001. Creating the gene ontology resource: design and implementation. Genome Res. 11: 1425-1433. See also http: //www.geneontology.org/

Tomita, M., Hashimoto, K., Takahashi, K., Shimizu, T., Matsuzaki, Y., Miyoshi, F., Saito, K., Tanida, S., Yugi, K., Venter, J.C., and Hutchison, C. 1999. E-CELL: Software environment for whole cell simulation. Bioinformatics, 15: 72-84. See also http: //www.e-cell.org/

Ziegler, B.P., and Weinberg, R. 1970. System theoretic analysis of model: computer simulation of a living cell. J. Theoretical Biology 29: 35-56.

From: *Bioinformatics and Genomes: Current Perspectives*
Edited by: Miguel A. Andrade

Chapter 8

Mapping Words For Genome Data Integration

Carolina Perez-Iratxeta and
Miguel A. Andrade

Abstract

In molecular databases, each entry appears commonly related to others, within the database and in different databases, through cross-linking associations. Moreover, each single entry is annotated, more or less systematically, using words from an ontology: a controlled and structured set of concepts. Knowledge can be gained by means of comparing the annotations of the items within or between databases. We present a very general scheme to derive existing associations between features or elements in databases. This heuristic approach can be applied to discover associations between the annotations of the entries. An application is presented consisting in a mapping from the MeSH (Medical Subject Headings) ontology of MEDLINE to the one of SWISS-PROT keywords. The mapping can be incorporated in a protocol for suggesting to the SWISS-PROT annotators keywords for new entries and, in some cases, correcting previous inconsistencies.

Introduction

During the year 2001, with the publication of the initial sequence and preliminary analysis of the human genome (International Human Genome Sequencing Consortium, 2001; Venter *et al.*, 2001), it was claimed that we entered in a new era for molecular biology, and indeed a significant amount of research will be directly inspired or intended by the analysis of genome sequence data. Whether we are in a new era or not, the fact is that a huge amount of genomic information has been obtained in recent years (Figure 1, taken from NCBI). At the present day, the complete genome sequence of about 60 organisms is already available, and approximately 300 genomes more are ongoing at different stages of completion.

Sequence data are being accumulated in gene and protein sequences databases. But sequence data are not the only important data and other pieces of molecular biology information are being intensively collected. For example, functional information regarding whole or partial genomes can be massively obtained through gene expression data analysis. Several consortia are working on increasing the number of 3D models for proteins via structural genomics projects (Edgar *et al.*, 2002; Bono *et al.*, 2002; Lo Conte *et al.*, 2002). Moreover, there is another very important source of information: research papers published in scientific journals. Currently more than 11 million papers are indexed in the MEDLINE database (see e.g. http://www.ncbi.nlm.nih.gov/PubMed/) and the number of published papers per year is actually increasing every year.

In order to make the accession to all these literature, functional, and structural information (as well as inferences based on protein families via sequence homology comparisons) easy and fruitful, all these data have to be integrated with sequence information and organized in databases.

The Molecular Database Essentials: Links And Words

Correlated with the increase on the amount of information, as expected there has been a large growth in the number and variety of the publicly available databases. A sound journal like *Nucleic Acids Research*, publishes a special issue at the beginning of January fully devoted to reports and descriptions of databases in the field of molecular biology. Significantly, the latest of those special issues contains around one hundred reports (*Nucleic Acids Res*. 2002 Vol. 30).

In molecular databases, often the items described are macromolecular sequences (polynucleotide sequences, proteins), and their properties. For

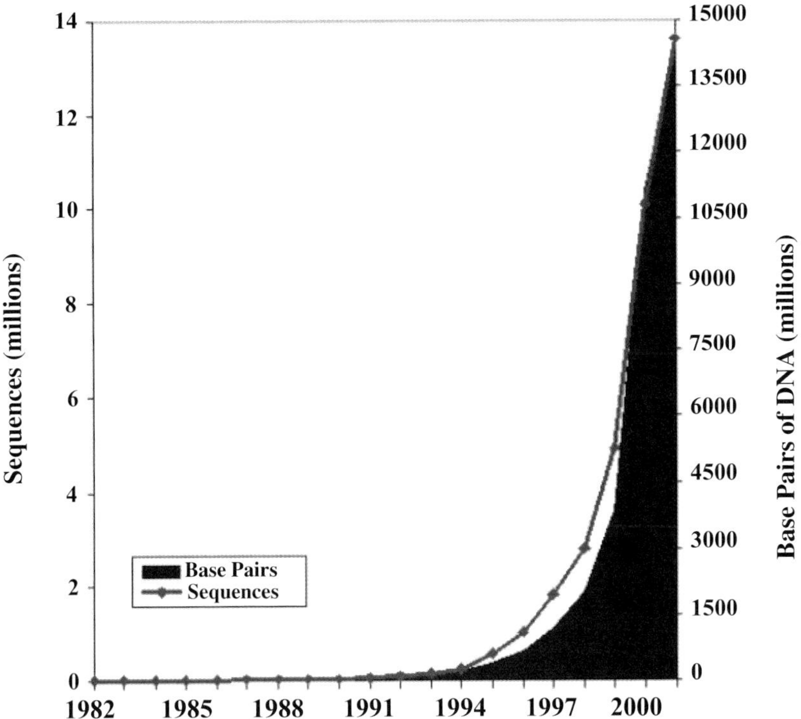

Figure 1. Growth of GeneBank in the recent past years (taken from NCBI).

example, for genes these properties can be mutations, genetic diseases, splice variants, expression data, etc; for proteins we can have motifs, domains, modifications, 3D structures, function, participation in complexes, metabolic pathways, signalling cascades, etc. Those properties can be the objects in other databases. In addition macromolecules are related by the biological processes producing them. For example a species contains a genome, which contains genes that may become translated into proteins that could be expressed in certain tissues, or under certain conditions, etc. Lastly, biological literature is the primary source of all this information and therefore can also be linked to it.

Accordingly, in molecular databases, each entry appears commonly related to others, within the database and in different databases, through cross-linking associations. The links enrich highly the information that is contained in a database as they represent the integration with another piece of information. Also links can be seen as a kind of 'variable' and hence handled by computers. Web interfaces allow the easy inclusion of hyperlinks in the database entries (and machineries designed to foster a collection of databases and their links have been developed, e.g. SRS, Etzold *et al.*, 1996). However, even with all the available technology, it is not unusual to get errors when following links across databases. If the links do not work properly a database looses a lot of its value.

Apart from links, there is another very important element in databases: the collection of words used to describe the data features. Usually, each single entry is annotated more or less systematically, using words (keywords) coming from restricted simple dictionaries or from ontologies. This annotation is most often performed manually by experts, the database curators. Database curators read the literature they consider to be relevant and select from the dictionary those keywords that best describe the subject. Keywords can give quick insights to database users and are useful for bioinformatics applications. In the case of the SWISS-PROT protein database, its entries possess an optional field labelled as KW (for keyword) which tries to summarize the main features of the protein with a few selected terms coming from a controlled vocabulary (Bairoch and Apweiler, 2000).

Ontologies or thesauri are more than controlled vocabularies. Indeed, they are preferred to unstructured sets of vocabularies. An ontology is a controlled and structured set of vocabulary representing the concepts and relations that are useful to describe objects and facts in a particular area of knowledge. Relationships between terms in ontologies are themselves meaningful and can be technically powerful, i.e., for browsing the database. The terms of an ontology are often connected by "broader-than," "narrower-than," and "related" links. These links show the relationship between related terms and provide a hierarchical structure that permits searching at various levels of specificity from narrower to broader. Thesauri are also known as "classification structures," and "ordering systems."

One advantage of such a formatted description is that it is easily managed by a computer. A database in which entries are documents written in natural language is very hard to treat with computational tools. One example of the latter is the On Line Mendelian Inherited in Man (OMIM) that is devoted to human genetic diseases (Hamosh *et al.*, 2002). This is a very well maintained database, where a lot of worthy information from the literature is read and digested by experts. But in fact most of this information cannot easily be exploited by computers since the entries are written in natural language directed to human interpretation.

One example of a structured vocabulary is the Medical SubHeadings terms, MeSH, from the National Medicine Library, NLM (http://www.nlm.nih.gov/mesh/meshhome.html). It is used to annotate each

MEDLINE abstract. Each paper that is recorded is associated with several MeSH terms, typically about a dozen, which summarizes the subject and features of the paper. The task is carried out manually -for each paper!- by an expert.

One example of a molecular biology ontology is Gene Ontology, GO (The Gene Ontology Consortium, 2001). The aim of this pioneering project (http://geneontology.org) is 'to produce a dynamic controlled vocabulary that can be applied to all organisms even as knowledge of gene and protein roles in cells is accumulating and changing'. GO is fairly structured under three organizing principles: 'molecular function', 'biological process', and 'cellular component', that allow meaningful descriptions and classifications of biological objects. Although it is not complete, just having a vocabulary that is both dynamical and structured and can be shared along different databases is already a big achievement.

An Example Of Data Integration From Literature And Proteins

As an example of how data integration could exploit the already stored information by means of controlled vocabularies and cross-linking, we present a practical application. It could be used for the functional annotation of proteins with keywords, and also for correcting some kind of internal inconsistencies within a database of such annotated proteins. The schema is very general and could be used to derive existing associations between features and elements in databases other than keywords.

Mapping Words Through Cross-Linking

The central idea is to put the literature data from MEDLINE via MeSH terms into contact with the protein function information from SWISS-PROT via its keywords. These contacts can be used to automatically generate suggestions to the annotator of keywords for a SWISS-PROT entry with the goal of easing the manual annotation task and preventing missing keywords i.e. those that logically should be present in the presence of another.

Krestchmann *et al.* (2001) generated a set of associations between SWISS-PROT keywords and properties of the sequences annotated with the keywords (namely, taxonomy of the organism where the protein is found, and presence of sequence patterns). Here, we construct a mapping between MeSH terms and SWISS-PROT. This mapping simply automatically associates a given MeSH term with one or more SWISS-PROT keywords.

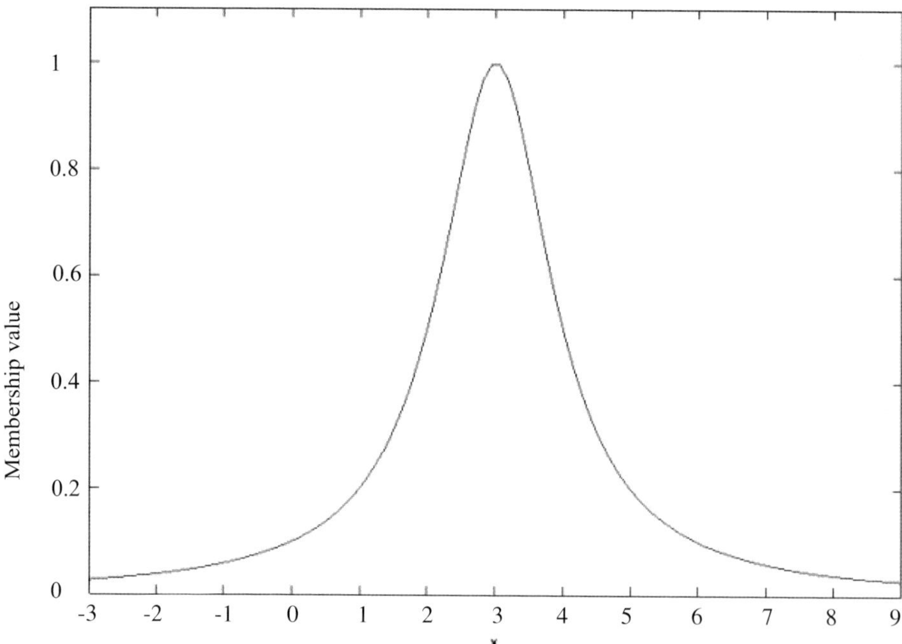

Figure 2. Graphical representation of the membership function of the fuzzy set 'real numbers close to 3' as defined in the text.

We can consider that two terms have to be associated through the mapping if they are related at some instance (i.e. 'Saccharomyces' and 'cerevisiae'). In order to formalise these relations between words and their strength, fuzzy binary relations are used (Zimmermann, 1985). In the next section we introduce this concept.

Fuzzy Binary Relations

A traditional set is a crisp set. One element of the universe either belongs to the set or does not. This is formally expressed by means of the characteristic function associated to the set. If one element belongs to the set, the value attached to it by the characteristic function is 1. If it does not belong to the set, the attached value is 0. In contrast, elements on fuzzy sets can have different degrees of membership of the set. The characteristic function, which now is called the membership function, can take real values within the interval [0,1].

For example, consider the fuzzy set 'real numbers close to 3'. The membership function of the set could be defined as the real valued function such that $f(x) = 1/[(x-3)^2+1]$, for every real x (see Figure 2). The function values increase fast when approaching 3, where the maximum is reached.

The above definition is suitable to model the implicit vagueness of a natural language expression like 'close to'.

A binary relation between two sets *A* and *B* is a subset of all the possible ordered pairs you can compose of the form (a,b), where a belongs to *A* and b belongs to *B* (the order is essential). Such a relation is a crisp set: a pair (a,b) belongs to the set if a and b are related, otherwise it does not belong. A fuzzy binary relation can be defined to model that elements are more or less related. This is useful in case we want to express different strengths in the relationship between elements. For example, to define the relation "X ressembles to Y" given a family photo album.

In the present case, MeSH terms and SWISS-PROT keywords are related. In a SWISS-PROT entry annotated with keywords there are also links to MEDLINE abstracts which in turn are annotated with MeSH terms. A mapping between these two classes of items would express the fact that some MeSH terms tend to occur often in abstracts that are linked to proteins that share some feature (expected to match some keywords) in common. For example, most of the MEDLINE abstracts indexed with the MeSH term 'Allosteric Regulation' are linked to entries in SWISS-PROT annotated with the keyword 'Allosteric enzyme'.

In general, for any SWISS-PROT entry we have on the one hand its keywords, and the other hand the MeSH terms of the MEDLINE abstracts linked from the entry. To define the fuzzy binary relation between a particular MeSH term m and a SWISS-PROT keyword sw, we have to estimate in some way the value of the membership function for the pair (m,sw). This is estimated by the number of SWISS-PROT entries where m and sw co-occur divided by the number of SWISS-PROT entries where m occurs. This is measuring the 'degree of inclusion' of m in sw.

Moreover, for each MeSH term m, the set of all its related SWISS-PROT keywords sw_i and on the distribution of the corresponding association values is considered. This particular distribution can tell us about the 'valuable information' (related to SWISS-PROT information) of finding it in an abstract. A score for each m, reflecting the information content of the MeSH term (informative value), is computed as the mean value of the distribution of log [assoc(m,sw_i)/fr(m)], where assoc(m,sw_i) is the association value of m to a SWISS-PROT keyword sw_i, and fr(m) is the frequency of the MeSH term m.

Assisting Manual Functional Annotation

Given an established SWISS-PROT entry or a new one, the MeSH terms of the linked abstracts are considered. As an example let us consider the SWISS-PROT entry ABA1_TRIAB (see Figure 3). This protein is the subunit 1 of

```
ID   ABA1_TRIAB     STANDARD;      PRT;   131 AA.
AC   P81111;
DT   15-JUL-1998 (Rel. 36, Created)
DT   15-JUL-1998 (Rel. 36, Last sequence update)
DT   15-JUL-1998 (Rel. 36, Last annotation update)
DE   Alboaggregin A subunit 1.
OS   Trimeresurus albolabris (White-lipped pit viper).
OC   Eukaryota; Metazoa; Chordata; Craniata; Vertebrata; Euteleostomi;
OC   Lepidosauria; Squamata; Scleroglossa; Serpentes; Colubroidea;
OC   Viperidae; Crotalinae; Trimeresurus.
OX   NCBI_TaxID=8765;
RN   [1]
RP   SEQUENCE.
RC   TISSUE=Venom;
RX   MEDLINE=98189535; PubMed=9531050;
RA   Kowalska M.A., Tan L., Holt J.C., Peng M., Karczewski J.,
RA   Calvete J.J., Niewiarowski S.;
RT   "Alboaggregins A and B. Structure and interaction with human
RT   platelets.";
RL   Thromb. Haemost. 79:609-613(1998).
CC   -!- FUNCTION: BINDS TO PLATELET GPIB/IX RECEPTOR SYSTEM AND STIMULATES
CC       AGGLUTINATION.
CC   -!- SUBUNIT: HETEROTETRAMER OF THE SUBUNITS 1, 2, 3 AND 4,
CC       DISULFIDE-LINKED.
CC   -!- SIMILARITY: TO OTHER MEMBERS OF THE C-TYPE LECTIN FAMILY.
DR   HSSP; P23806; 1IXX.
DR   InterPro; IPR001304; lectin_c.
DR   Pfam; PF00059; lectin_c; 1.
DR   SMART; SM00034; CLECT; 1.
DR   PROSITE; PS00615; C_TYPE_LECTIN_1; 1.
DR   PROSITE; PS50041; C_TYPE_LECTIN_2; 1.
KW   Venom; Lectin.
FT   DOMAIN         1     129       C-TYPE LECTIN (LONG FORM).
FT   DISULFID       2      13       BY SIMILARITY.
FT   DISULFID      30     127       BY SIMILARITY.
FT   DISULFID     102     119       BY SIMILARITY.
SQ   SEQUENCE   131 AA;  15427 MW;  B3569F5BF91F6624 CRC64;
     DCPSDWSSYD QYCYRVFKRI QTWEDAERFC SEQANDGHLV SIESAGEADF VTQLVSENIR
     SEKHYVWIGL RVQGKGQQCS SEWSDGSSVH YDNLQENKTR KCYGLEKRAE FRTWSNVYCG
     HEYPFVCKFX R
//
```

Figure 3. SWISS-PROT entry for the protein ABA1_TRIAB.

the alboaggregin A from the viper species *Trimeresurus albolabris* (Kowalska *et al.*, 1998). This entry is poorly annotated with SWISS-PROT keywords, the only ones assotiated being 'Venom' and 'Lectin'. One MEDLINE abstract is linked to ABA1_TRIAB, with identifier 98189535. The list of MeSH terms for 98189535 is displayed in Table 1.

Among them we can find several terms that are general and as a consequence they have weak association to SWISS-PROT keywords, i.e.

Table 1. List of MeSH terms indexed in the MEDLINE entry 98189535. Numbers on the right column refer to the informative value.

Amino Acid Sequence	0.15
Blood Platelets/*drug effects/pathology	2.02
Crotalid Venoms/*pharmacology	2.79
Human	0.48
Molecular Sequence Data	0.018
Platelet Aggregation/*drug effects	3.21
Platelet Aggregation Inhibitors/*pharmacology	3.37
Sequence Alignment	0.058
Structure-Activity Relationship	0.61
Support, U.S. Gov't, P.H.S.	-

'Amino Acid Sequence'. Consequently, their respective informative value is very low (see Table 2). We also find *informative terms* such as 'Platelet Aggregation' and 'Platelet Aggregation Inhibitors'. These two terms point to the same SWISS-PROT keywords (see Table 2). From here, an automatically generated suggestion can be made to the annotator, indicating that 'Platelet', 'Blood coagulation', and 'Cell adhesion' should be considered. Actually, these keywords are correctly matching the function of the protein in the present case.

Correcting Internal Inconsistencies

Also, a set of general rules can be generated based on the association within SWISS-PROT keywords. Consider a new mapping derived in an analogous way but relating SWISS-PROT keywords with themselves. Again, those

Table 2. SWISS-PROT keywords more strongly associated with the 'Platelet Aggregation' and 'Platelet Aggregation Inhibitors' MeSH terms. The association values are multiplied by 100. Note that both MeSH terms are pointing to a similar set of keywords.

Platelet Aggregation Inhibitors		**Platelet Aggregation**	
Platelet	6.44	Platelet	6.41
Blood coagulation	5.44	Blood coagulation	5.42
Cell adhesion	5.64	Cell adhesion	4.57
Venom	4.53	Venom	4.61
Lectin	3.69	Lectin	3.67
3D-structure	1.51	Hydrolase	0.77
Hydrolase	0.32	Signal	0.39
Signal	-0.01		

Table 3. Top of a list of pointers to potential inconsistencies in particular SWISS-PROT entries based on the associations computed between SWISS-PROT keywords. The list is ranked by likelihood. The last two would be rejected by the curator, as it is discussed in the text.

Entry	Inconsistency	Confidence
oar_phopy	Transmembrane is MISSING because of G-protein_coupled_receptor	0.9991
hm1d_droan	Nuclear_protein is MISSING because of Homeobox	0.9986
gcn2_yeast	Protein_biosynthesis is MISSING because of Aminoacyl-tRNA_synthetase	0.9985
gcn2_yeast	Ligase is MISSING because of Aminoacyl-tRNA_synthetase	0.9985
luci_renre	Lyase is MISSING because of Decarboxylase	0.9976
tnfa_caphi	Transmembrane is MISSING because of Signal-anchor	0.9976
sn14_yeast	Protein_biosynthesis is MISSING because of Elongation_factor	0.9973
cp51_wheat	Transferase is MISSING because of Methyltransferase	0.9972
cp51_sorbi	Transferase is MISSING because of Methyltransferase	0.9972
if3_bacsu	Protein_biosynthesis is MISSING because of Initiation_factor	0.9971
acsd_mooth	Lyase is MISSING because of Carbon_dioxide_fixation	0.9967
acsc_mooth	Lyase is MISSING because of Carbon_dioxide_fixation	0.9967

keywords that tend to occur together in entries may be in some way related. A set of rules can be derived from the mapping. For example, we find that the pair ('Homeobox', 'Nuclear Protein') has a high value within the relation. Then we can derive the following rule for annotation: "If Homeobox is present, Nuclear Protein should be also present". An automatic protocol can be easily implemented for checking inconsistencies in the database, i.e., finding which entries are not fulfilling a particular rule and, automatically pointing them to the curator for checking or revision (see Table 3). Notice that such a set of rules would palliate the fact that SWISS-PROT keywords do not constitute a thesaurus.

We want to stress the fact that these rules are automatically generated without using any biological knowledge, and that they are context dependent. For that reason they are not necessarily true. An extreme example: a total of 600 entries are annotated in SWISS-PROT with the keyword 'Carbon dioxide fixation'. All of them but two are also annotated as 'Lyase'. The two exceptions are the entries ACSD_MOOTH and ACSC_MOOTH, corrinoid/Fe-S proteins, which act as a methyl group carrier, and are certainly not lyases (see bottom of Table 3).

Conclusion

In order to get the most of the raw data derived from genome sequencing projects, it is necessary to provide coherent and clean data collections with the appropriate and maximally extended set of links. The parallel evolution of many databases, the continuous development of new ones, and the increment in data production as sequencing and new massive approaches gather data from complete genomes and living organisms, makes it essential to use computational tools and controlled vocabularies to manage the generation and maintenance of the links between and within databases.

In this chapter we have presented one possibility for improving the coherence of features associated with the entries of a database. An appropriate functional annotation for gene sequences by means of ontologies, and cross-linking between databases were two starting essential points. From there, data mining techniques can be applied to knowledge extraction or, as it is suggested above, to develop tools for helping manual annotation and curation.

Acknowledgments

Thanks to the NIH/NLM for maintaining and providing MeSH, and to the SWISS-PROT teams (in the European Bioinformatics Institute, EBI, Hinxton, and in the Swiss Institute of Bioinformatics, Geneve) for maintaining SWISS-PROT.

References

Bairoch, A., and Apweiler, R. 2000. The SWISS-PROT protein sequence database and its supplement TrEMBL in 2000. Nucleic Acids Res. 28: 45-48.

Bono, H., Kasukawa, T., Hayashizaki, Y., and Okazaki, Y. 2002. READ: RIKEN Expression Array Database. Nucleic Acids Res. 30: 211-213.

Edgar, R., Domrachev, M., and Lash, A. E. 2002. Gene Expression Omnibus: NCBI gene expression and hybridization array data repository. Nucleic Acids Res. 30: 207-210.

Etzold, T., Ulyanov, A. and Argos, P. 1996. SRS: information retrieval system for molecular biology data banks Methods Enzymol 266: 114-128.

Hamosh, A., Scott, A.F., Amberger, J., Bocchini, C., Valle, D., and McKusick, V. A. 2002. Online Mendelian Inheritance in Man (OMIM), a knowledgebase of human genes and genetic disorders. Nucleic Acids Res. 30: 52-55.

International Human Genome Sequencing Consortium. 2001. Initial sequencing and analysis of the human genome. Nature 409: 860-921.

Kretschmann, E., Fleischmann, W., and Apweiler, R. 2001. Automatic rule generation for protein annotation with the C4.5 data mining algorithm applied on SWISS-PROT. Bioinformatics 17:920-926.

Kowalska, M.A., Tan, L., Holt, J.C., Peng, M., Karczewski, J., Calvete, J.J., and Niewiarowski, S. 1998. Alboaggregins A and B, Structure and interaction with human platelets. Thromb. Haemost. 79: 609-613.

Lo Conte, L, Brenner, S.E., Hubbard, T.J.P., Chothia, C., and Murzin, A.G. 2002. SCOP database in 2002: refinements accommodate structural genomics. Nucleic Acids Res. 30: 264-267.

The Gene Ontology Consortium. 2001.Creating the gene ontology resource: design and implementation. Genome Res. 11: 1425-1433.

Venter, J.C., Adams, M.D., Myers, E.W., Li, P.W., Mural, R.J., Sutton, G.G., Smith, H.O. *et al.* 2001. The sequence of the human genome. Science 291: 1304-1351.

Zimmermann, H.J. 1985. Fuzzy set theory and its applications. Kluwer Academics Publishers, Boston.

From: *Bioinformatics and Genomes: Current Perspectives*
Edited by: Miguel A. Andrade

Chapter 9

Structural Genomics and Structural Bioinformatics

Patrick Aloy, Baldomero Oliva
and Robert B. Russell

Abstract

During the last years, we have seen how molecular biology has gone from sequencing a single gene to sequencing an entire eukaryotic genome. The large-scale sequencing projects are producing more and more sequences and the sequence databases are growing exponentially. However, full understanding of the biological role of these proteins will require knowledge of their structure and function. The different structural genomics projects were born as an attempt to tackle the problem of protein structure determination for complete genomes (proteomes). Experimental and computational approaches are needed to fulfil the expectative of assigning structures to proteomes. The experimental level (*Structural Genomics*) includes a large-scale cloning, expression, purification and structure determination of the proteins. The computational level (*Structural Bioinformatics*) includes protein target selection, structure interpretation, comparative modelling and, finally, prediction of protein function. In this chapter, our intention is to take you through all these different aspects of Structural Genomics and to give you a rough idea of the state-of-the-art techniques and discuss the foreseen improvements and limitations.

Structural Genomics

Target Selection

The first step in any high-throughput structural genomics initiative is the rational selection of the proteins whose three-dimensional structures have to be characterised. In other words, to decide which structures, from which species and in which order have to be solved. Such a process is known as target selection, the target being a region of a protein to be studied by crystallography or NMR.

Since it is not feasible, at short term, to solve the structures for all proteins in all organisms and we want to get a reasonable coverage of the fold and superfamily space in the shortest time possible, it is essential to choose very carefully the targets and to internationally coordinate efforts in order to obtain the maximal gain from all the different initiatives. Target selection is mainly a computational process and consists on restricting candidate proteins to those that are tractable and of unknown structure, and prioritising according to expected interest and accessibility. Steven Brenner divided the target selection process into four sequential steps: realm identification, family exclusion, family prioritisation, and protein region selection (Brenner, 2000).

The realm identification is the selection of the general universe of interest for a particular structural genomics project. Some groups are focusing on a particular organism, such as a hyperthermophile. In this case the realm of interest is obvious: all the proteins in the organism. However, other initiatives are focusing on the proteins related to a particular disease or on obtaining a better understanding of spread protein families such as signalling proteins or on a specific metabolic pathway. In these cases, the realm identification consists of defining which organisms or which kind of proteins are the system of interest.

Family exclusion includes the exclusion of proteins that can be modelled by homology and those that can prove very difficult for protein determination. As the main goal of structural genomics is to cover the fold space as fast as possible, all those proteins that share more than 30% of sequence identity with a known structure should be removed from the target list as it is clear that they will present the same fold as their homologues and their structure can be obtained by means of homology modelling (see *Comparative Modelling* below).

Some proteins can contain regions with certain features known to enormously difficult the crystallising procedure. A clear example is the long low-complexity regions (with only one or two different amino acid types; Wootton, 1994). They might play different roles such as linkers or formation of unusual regular structures (e.g. poly Q, Perutz, 1999), or unstructured regions before binding. What is clear is that they constitute mobile regions difficult to crystallise and therefore they should be removed from an initial list of priority targets. A different case is that of transmembrane (TM) proteins. It has been estimated that 30% of proteins are TM including those of most interest for biomedical purposes. These TM proteins are very difficult to crystallise due to their hydrophobic environment and should also be removed from the initial list. However, this does not mean that these protein structures cannot be determined. For example, G-protein coupled receptors contain little more than TM segments, and yet could be modelled in some cases on the structure of human rhodopsin, or on other transmembrane proteins whose structure has already been determined.

Family prioritisation is ultimately unnecessary as, by definition of structural genomics, all the families that have survived the exclusion process have to be tackled. However, there can be proteins that are more interesting for a given Structural Genomics initiative and that are worth prioritising. Examples can be proteins taxonomically dispersed in all the realms (ancient and conserved and represent important biological processes), those that will permit more proteins to be modelled or on the contrary those that are unique (ORFans). Or those found in *C.elegans* and not in *S.cerevisiae* and therefore specific for higher eukaryotes.

Multidomain proteins are more difficult to express in *E.coli* and to crystallise due to their larger size and to the mobile regions linking the different domains. So, when one attempts to solve the three dimensional structure of one of these proteins, it is usually better to break a long protein into its component domains and solve them separately. The structural domains are compact, able to fold by themselves and usually associated with discrete aspects of molecular function.

Finally, other aspects of the target selection process include the screening of orthologues for optimal conditions (e.g. more stable proteins, appropriate pH or a larger number of methionines suitable for multiple anomalous dispersion "MAD" experiments).

Protein Production

The different Structural Genomics projects going on will need to produce thousands of proteins during the next few years to feed the structural biologist needs. Hence, the development of efficient systems for protein cloning, expression and purification will be extremely valuable.

Cloning is the first step in protein production. With the current techniques, it is almost recombinant DNA routine to produce expression vectors for genes without introns. For these proteins, the challenge for structural genomics will be to completely automate the process in order that a few of people can create thousands of clones per year. The picture is very different when we deal with eukaryotic, intron containing, genes. For cloning these genes, the construction of expression vectors is not straight forward and one has first to identify and remove the introns from the gene sequence. Therefore, at this stage, the structural genomics initiatives, together with other molecular biologists, should focus on the construction of libraries of full-length cDNA that can afterwards be used in the same automated pipeline developed for the prokaryotic genes. As the ultimate goal will be to develop a fully automated strategy for protein production, another important aspect to take into account when designing the expression vectors is the choice of an affinity tag in order to simplify the purification step. Up to date, the introduction of polyhistidine tails and the use of glutathione S-transferase (GST) as fusion protein are the most common strategies.

Once the protein of interest has been cloned, we will need to express it for its final structure determination. If our protein expresses well in soluble form in bacteria, the expression step will represent no challenge at all. Proteins expressed in *E.coli* can be purified easily and will facilitate the crystallisation process or give good NMR spectra. The *E.coli* system permits to use the bacterial metabolism to replace the methionines by seleno-methionines in our protein to be used in MAD crystallography techniques, or to label our protein with ^{15}N and ^{13}C for NMR studies. There are also well-established methods for coping with different codon usage in *E.coli*, like supplementing the cells with tRNA genes coding for the low frequency codons. Unfortunately, only 15 to 20% of the proteins of interest will behave like this (Edwards *et al.*, 2000). The vast majority of eukaryotic proteins are formed by multiple domains that, in addition, require either post-translational modifications or other proteins or cofactors to fold correctly, and they will be expressed in insoluble form, if expressed at all, in bacterial systems. All these factors make the expression of eukaryotic proteins in bacterial systems a difficult task. Unfortunately, most of the new advances described above have been developed in *E.coli* and their direct extrapolation has to be further interrogated.

For such proteins, there exist a handful of strategies that can be used in order to express enough protein to proceed with the structural analyses. One of the most used is to split the proteins into structural domains (see *Target Selection* above) that can be better expressed in *E.coli* than multidomain proteins and will crystallise more easily. Another strategy is to clone and express several orthologues from different organisms and then choose the most soluble. They will all present the same 3D structure and catalytic mechanism, but subtle changes in the amino acid sequence can affect the solubility, and therefore the chances of being correctly expressed, in bacterial systems. Some proteins can be poorly expressed because of the lack of an essential cofactor. However, when dealing with new proteins, this situation is almost impossible to predict. We might try the chemical proteomics approach, which consists of screening the expression of our protein of interest in the presence of several potential small cofactors (i.e., ATP) in order to increase it expression. If none of these strategies work, we will have to try with a different expression system (i.e., yeast, human cells, etc.), usually more expensive and difficult to manipulate.

The purification is the last step in the protein production pipeline. New developments will be needed in this area as the current purification techniques require considerable expert knowledge and manual intervention and are not suited to fulfil the structural genomics requirements. However, when the protein of interest is expressed as a recombinant protein with an affinity tag, such as a hexahistidine tail or GST, the process is much easier. In these cases, an affinity chromatography, with final removal of the tag, usually provides a major step of purification. If the structure of the protein is to be determined by NMR, there is no need to remove the histidine tail, as it does not interfere with the spectrum. For crystallography, removal of the tag and additional purification steps might be required, as the probability of obtaining crystals increases with the purity of the sample.

Structure Determination

Once our protein of interest has been expressed arises the question of which method to use for its structure determination: X-ray crystallography or Nuclear Magnetic Resonance?

X-ray Crystallography

The X-ray crystallography takes advantage of the interaction of X-rays with electrons in order to determine the structure of macromolecules at nearly atomic model. As a single molecule is a very weak diffractor of electrons, we have to obtain crystals to amplify the diffraction signal. These crystals contain many molecules (in the order of 10^{15}) repeated in an identical orientation in a regular arrangement so that they diffract identically.

We have to go through a series of sequential steps, the first of which is obtaining the crystals. To get crystals of a macromolecule, we need to supersaturate the sample. The most common technique is to use a hanging drop containing the concentrated protein in a few micro-litres droplet and to slowly dehydrate it by vapor diffusion against a reservoir of the precipitant. The concentration needed to obtain crystals is supraphysiological and depends on the solubility of the protein, but usually ranges between 5 and 50 mg/ml. A wide variety of parameters can affect the diffraction-quality of the crystals. They include pH, precipitant (salts, polyethylene glycols, organic solvents, etc.), concentration, temperature, ions, additives, etc. Nowadays, we are still unable to predict the best crystallisation conditions for a given protein from its sequence. Hence, in practice, the strategy is to first solve the nucleation problem and then the growth problem. To first obtain crystals we use screening protocols for checking a wide range of crystallization conditions with a relatively small amount of protein. The favorite ones are the Hampton screens, where we screen about 200 different conditions at different temperatures. Crystals on average appear between a few days and a few weeks. To then grow the crystals in a suitable form and size, we set up more rational finer screenings of the parameters. New generation robots have been developed for the automated screening of such conditions.

As soon as a well diffracting crystal is available, the next step is the data collection. This is the last experimental step in structure determination. X-rays are obtained by bombarding a Cu target with electrons produced by a heated filament and accelerated by an electric field. The target is continuously rotating to dissipate heat and its emission is filtered so that monochromatic $K\alpha$ radiation is selected at a fixed wavelength. In a synchrotron, electromagnetic radiation is produced by accelerated charged particles moving at relativistic speed. Electrons or positrons with high energy (GeV) move at nearly the speed of light along a circular path, so that they acquire angular acceleration even if their linear velocity is constant.

Solving the phases is the next step once the diffraction data has been collected. Since with a diffraction experiment we can only directly derive the amplitude of each diffraction maximum, we have to estimate its phase by indirect means. The most common phasing methods are molecular replacement (MR) in cases where a sufficiently homologous structure is known, or multiple isomorphous replacement (MIR) and multiple anomalous dispersion (MAD) where no similar structure is known.

If the structure of a homologous protein is known, we can use its coordinates to calculate an initial estimate of the phases. This method is known as Molecular Replacement (MR). In this approach the homologous probe structure is fit into the unit cell of the unknown structure. The success of the technique depends largely on the structural identity between the probe and the unknown protein. With the rapidly increasing number of three-dimensional structures determined, the application of MR as a general procedure to solve the phasing problem is expected to be a reality in the near future.

The Multiple Isomorphous Replacement (MIR) method relies on the changes in intensities caused by a heavy atom bound to the crystalline protein molecule (heavy-atom derivative) as compared to the parent (native) diffraction pattern. Estimates of the phases can be derived from the differences between the amplitudes of the derivative and native diffraction data. Heavy-atom derivatives are prepared by soaking the native crystals in a buffer containing a heavy-metal compound (mercury, platinum, uranium, etc.), or by co-crystallizing a protein with it.

Multiple Anomalous Dispersion (MAD) exploits the changes in the diffraction pattern obtained when using different wavelengths around the absorption edge of certain atoms (anomalous scatterer) in the crystal. Different anomalous signals are collected by carefully choosing different wavelengths around the absorption edge of the compound. This experiment can therefore only be carried out with tunable synchrotron radiation.

The lasts steps are the model building and refinement. Once the phases are available, the electron density map can be calculated and interpreted by fitting the polypeptide backbone and the side chains into the electron density map. At this step is where the importance of the resolution data is most important. Beyond 3.5 Å resolution it is difficult to recognize β-sheets, and beyond 5 Å α-helices. In the case of high-resolution maps, computer programs have been developed recently to do model building automatically. X-ray crystal diffraction cannot resolve the position of hydrogen atoms apart from the few cases where the data extend beyond 1.2Å resolution.

The purpose of refinement and rebuilding is to detect and fix the errors in order to obtain the best possible final model. Refinement is used to change the model such that it fits the experimental data best. Also in the initial stages molecular dynamics can be used in order to avoid stereochemical bumps. Cycles of refinements will eventually reach convergence, after which we need to manually change and improve the model interactively on the graphics.

Nuclear Magnetic Resonance

Nuclear magnetic resonance (NMR) spectroscopy uses radio frequency radiation to induce transitions between different nuclear spin states of samples in a magnetic field. The utility of NMR for structural characterisation arises because different atoms in a molecule experience slightly different magnetic fields, and therefore transitions, at slightly different resonance frequencies in a NMR spectra. Furthermore, splitting of the spectra lines arise due to interactions between different nuclei, which provides information about the proximity of different atoms in a molecule.

To determine the structure of our protein of interest by NMR spectroscopy, the first thing we will have to do is to enrich the sample solution with ^{13}C and ^{15}N isotopes to resolve the spectral overlap in ^{1}H NMR. Currently, with the advances in protein expression in bacterial systems (see above *Protein Production*) such a labelling is a routine process. The next step is to put the sample into the magnet and record a set of multidimensional NMR experiments that will provide the data to obtain the NMR spectra. Once we have got the spectra, we have to assign first the shifts that correspond to contacts between residues close in sequence (sequence and secondary structure of the protein) and then the ones that are conformation dependent. Nowadays, the long-range contacts have to be manually assigned by determining the NOEs, scalar, and dipolar couplings. All these measures provide the information necessary to define the distance constraints. The final structure(s) is obtained by building up a model that satisfies as many of the derived distance constraints as possible. Finally, energy optimisation and few steps of molecular dynamics can be applied to refine the model(s).

New technical advances in the field, like the high-field magnets, the use of cryogenic probes or partial deuteration, will reduce considerably the time of data collection and improve its quality. In addition, the use of the TROSY (Transverse Relaxation Optimised Spectroscopy), that takes profit from slowly relaxing transitions, can provide a qualitative advance by enhancing the sensitivity for larger proteins (30-50 kDa).

In a Structural Genomics context, NMR can also be used to explore the whether a specific construct is going to be properly folded before any structure determination attempt. This can be easily done by calculating the correlation between ^{1}H 1D and ^{15}N or ^{13}C 2D NMR spectra. As we have already seen, the isotopic labelling is a quite easy and cheap procedure and therefore suitable for application in a high-throughput pipeline. It has also been proposed that there could be a correlation between this “foldedness” criteria and crystallisation conditions (Montelione *et al.*, 2000).

However, one of the most significant goals that NMR spectroscopy can achieve within the frame of Structural Genomics is the characterisation of structure-function relationships. When attempting functional studies (e.g. small molecules binding or presence of cofactors) differences in the chemical shifts of some residues are a good indicator for their implication in the binding. With such a technique, we can screen for many different ligands and apply it to drug design of ligand-binding epitopes.

The most important handicaps for NMR spectroscopy are the size limitation and the heavy instrumentation required. Currently, only proteins of less than ~175 residues can be tackled by this technique and often the atomic level obtained is lower than for structures solved by X-ray crystallography. Another problem is the lack of automated methods for interpreting NMR spectra (the shifts have to be assigned manually) and the difficulty in assessing the statistical correctness of the structure. However, it also presents clear advantages, the most obvious of which is that we do not require crystals of the protein for structure determination. Another advantage is that the studies are performed in aqueous solution, close to physiological conditions, and therefore structure-function relationships can be better studied. For instance, changing the conditions under which the experiment is performed, such as pH, may give us biophysical information very valuable for designing further experiments.

Although the two methodologies are clearly complementary, at the initial stages of Structural Genomics, small proteins will probably be tackled by NMR spectroscopy, whereas X-ray crystallography will be the choice for larger proteins and protein complexes.

Structural Bioinformatics

Interpreting Three-Dimensional Structures

Structure Comparison

Protein structures can be similar even when no similarity in sequence is detectable. Methods for comparing structures have been developed that allow one to find such similarities (e.g. Taylor and Orengo, 1989; Holm and Sander, 1996b; Russell and Barton, 1992). The idea is that one uses the three-dimensional coordinate data to find an alignment of protein sequences and an associated superimposition that can be seen using molecular graphics programs.

Protein Structure Classifications

Several protein structure classification schemes have become available via the internet over the last five years. The relative merits of each are discussed below. Perhaps the most important general comment is that none of the classifications give a complete picture. Since all have different strengths, it is best to consider all of them when attempting to put a structure in the correct context.

SCOP: Structural Classification of Proteins

SCOP is maintained by Murzin *et al.* (1995) in Cambridge, and is an entirely manual classification. Proteins are generally divided into functional domains, and these are grouped into a hierarchy consisting of class, fold, superfamily, family, protein, and species. Proteins are put into the same fold if they have a similar core, which is decided by manual analysis. The fold definitions in SCOP are more stringent than the other classifications, meaning that there are often structural similarities between "different" folds. The subdivisions within each fold (superfamily, family, protein, and species) group proteins according to their degree of homology.

The great strength of SCOP is the very careful assignment of evolutionary relationships, even in the absence of sequence similarity. Proteins in the same fold, but in different superfamilies, are lacking in evidence for a common ancestor (analogous folds – possibly convergently evolved?); those in the same fold and the same superfamily show some evidence of a common ancestor, which is often based on the features discussed above.

CATH: Class Architecture Topology Homology

CATH is maintained by Orengo *et al.* (1997) at University College, London. The classification is partly automated and partly manual, although they work towards a mostly automated system. They classify proteins according to a hierarchy that is similar to SCOP, though with some important differences. The Class (C) layer is similar to that of SCOP. The Architecture (A) layer, a unique feature of CATH, is an intermediate between class (C) and fold (or topology, T, in CATH). Protein architecture defines the orientation of the secondary structures composing the fold, independent of the connectivity or direction of secondary structure elements. For example, all protein domains containing a β-barrel are placed in a single architecture regardless of the number of strands forming the barrel, or the connectivity. The extra level of the hierarchy makes browsing classifications somewhat easier, and it makes structure space more continuous than in some of the other classifications. Architecture also encapsulates many of the descriptions often given with newly solved three-dimensional structures (e.g. the protein contains a β barrel structure). CATH provides excellent peripheral information for every protein structure in the database. Detailed graphical information as to structural motifs (Laskowski *et al.,* 1998) bound ligands (Wallace *et al.,* 1995) and cross references to many other data sources are all available.

FSSP: Families of Structural Similar Proteins

FSSP is provided by Holm and Sander (1996b) at the European Bioinformatics Institute (Hinxton, UK). Rather than a classification, FSSP is a list of protein structural neighbours. After each update of the PDB (the Protein Data Bank of protein structures), each new protein is compared to all others using sequence and structure comparison methods. Thus for each PDB entry, one can obtain a list of sequence and structure neighbours (the results of a search). There are no discrete boundaries discerning similarity from dissimilarity. Frequently, proteins that are not grouped in other classifications, e.g. SCOP or CATH, are structural neighbours within FSSP, reflecting weaker matches that may nevertheless represent biologically meaningful examples. Multi-domain proteins are compared as a whole to the database, meaning that it may be difficult to see similarities involving a rare domain when it is connected to one occurring frequently. A comparatively new feature is the DALI Domain Dictionary (Holm and Sander, 1998), now permitting comparisons between automatically detected domains.

Superfamilies: Does Structural Similarity Imply Homology?

It is probably impossible to state whether all proteins adopting a similar fold descend from a similar common ancestor (i.e., related through divergence). For many proteins with similar folds, sequence, structure or functional arguments suggest divergence from a common ancestor; for others, no such conclusion can be drawn. Hence some classifications distinguish between similarities that are due to divergent evolution, and those that may not be. It is clear that many homologous proteins have simply diverged beyond the point where sequence similarity can be detected. The term *superfamily* is often used in structure classification to refer to groups of proteins that appear to be homologous, even in the absence of significant sequence similarity.

Proteins with the same fold that are not thought to share a common ancestor are often referred to as *analogues* (to distinguish them from homologues), and are sometimes thought to be the result of convergence to a stable structure. Although there is little hard evidence, there are some arguments that favour such convergence processes. For instance, the number of proteins sampled during evolutionary time is vast, despite an estimated low number of possible folds (Chothia, 1992; Blundell and Johnson, 1993; Orengo *et al.*, 1994), which may be due to restrictions on protein architecture. How can homology be inferred when sequence identity is un-detectable? A survey of recent literature shows that one or more of the following features are often used to deduce a common ancestor (i.e., assign a common superfamily) given a pair of similar 3D structures:

1. Above a certain level of structural similarity, even if sequence similarity is insignificant, one can be largely confident of a homologous relationship (Orengo *et al.,* 1994; Holm and Sander, 1997).
2. The conservation of unusual structural features, sometimes outside the common core secondary structure elements. These features include functionally important turn conformations (Swindells, 1995), left-handed βαβ units (Murzin *et al.*, 1995) or others (e.g. Murzin, 1993).
3. Low but significant sequence identity as calculated after structure superimposition (i.e., the identity from the structure-based alignment Russell *et al.,* 1997). See reference Murzin (1993) for guide (illustrated by example) to how to calculate an associated statistical significance. Identities from a structure-based alignment of >12% are more likely to indicate a remote homology. Note also that structure similarities may confirm marginally significant sequence similarities seen prior to 3D structure determination.
4. The presence of key active site residues, even in the absence of global sequence similarity. This is most often applicable to enzymes (for examples see: Holm and Sander, 1995, 1997; Artymiuk *et al.,* 1997).
5. Sequence similarity bridges, or *transitivity*. Even though two sequences may not be significantly similar to one another, inspection of homologues found in sequence searches with each sequence may reveal a "link" or "sequence bridge" linking the two sequences via significant sequence similarities (e.g. Holm and Sander, 1996a, 1996b; Park *et al.,* 1997). In other words, if domain A is significantly similar to domain B and domain C is significantly similar to domain B, then domains A and C can generally be deemed homologous.

It is worth noticing that the structure with the highest degree of structural similarity to a probe structure may not necessarily be the best candidate for superfamily or functional similarity. This can be partly due to limitations in the structure comparison method.

Recently, we used a measure of sequence identity after structural superposition, defined as P_{3D}-values, to link the sequence space by means of structure (Aloy *et al.*, 2002). We applied the described methodology in a large-scale study, using the SCOP structural database to identify homologies not detectable by sequence methods alone and to merge different Pfam (Bateman *et al.*, 2000) and SMART (Schultz *et al.*, 2000) sequence domains (see Figure 1).

This study has gone some way towards quantifying how structural genomics projects will gradually link sequence families together. A protocol

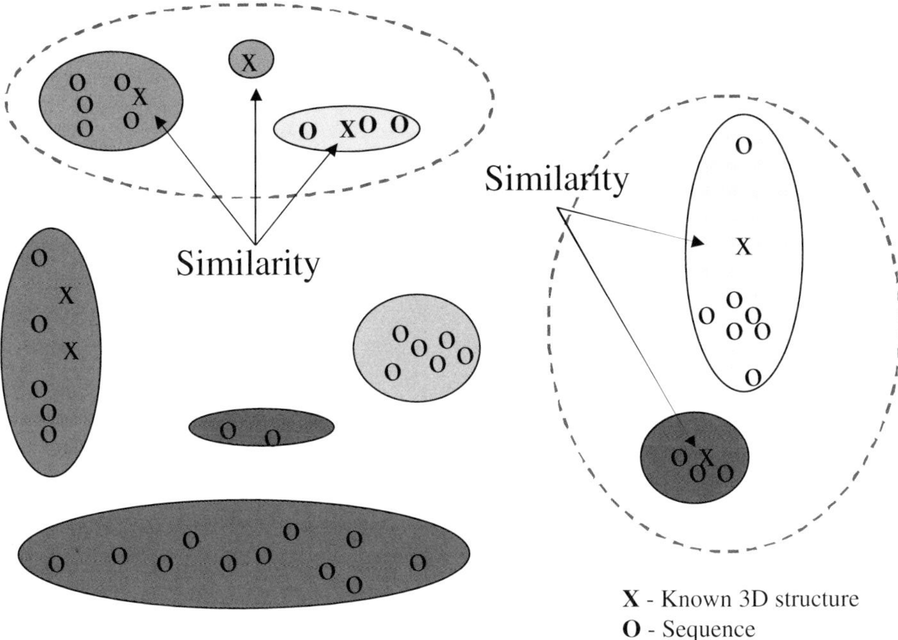

Figure 1. Structural similarity to link sequence space. Sequence similarity after structural superimposition is one of the best methods to unveil hidden evolutionary relationships between protein families not detectable by sequence-based methods. Protein sequences (O) are merged in the same family (continuous lines) when evidences of sequence homology are detected. However, protein families annotated as different by sequence-based methods can be fused (dashed lines) if a representative structure from each family (X) is available and structure-based comparisons show significant results.

like this will permit evolutionary and functional similarities to be uncovered automatically as the number of known structures and sequences continues to increase. In this context of structural genomics, where a protein three-dimensional structure can be known before its function, the ability to place a new structure in the correct evolutionary context is currently the best method for predicting details regarding molecular function.

Specific Example of Using Structure to Link the Sequence Space: A New Potential Superfamily Within the Ferredoxin-like Fold

We applied the methodology described above and found several potential links between different superfamilies within the ferredoxin-like fold. This fold comprises a repeat of a split αβα motif that forms an anti-parallel β-sheet flanked on one side by two α-helices. It is one of the most populated in SCOP, with 31 different superfamilies and also performs a large number of

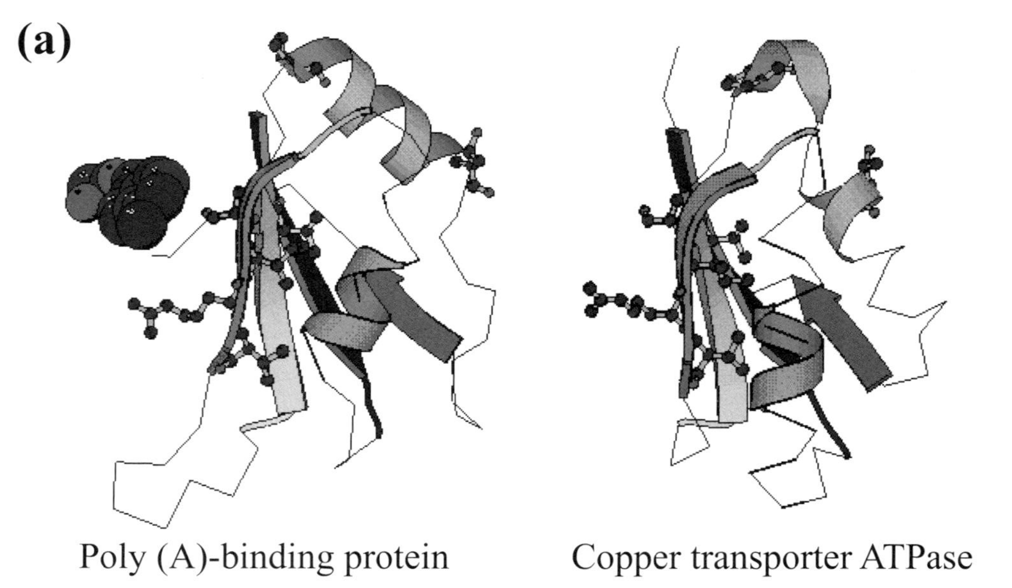
(a)
Poly (A)-binding protein
Copper transporter ATPase

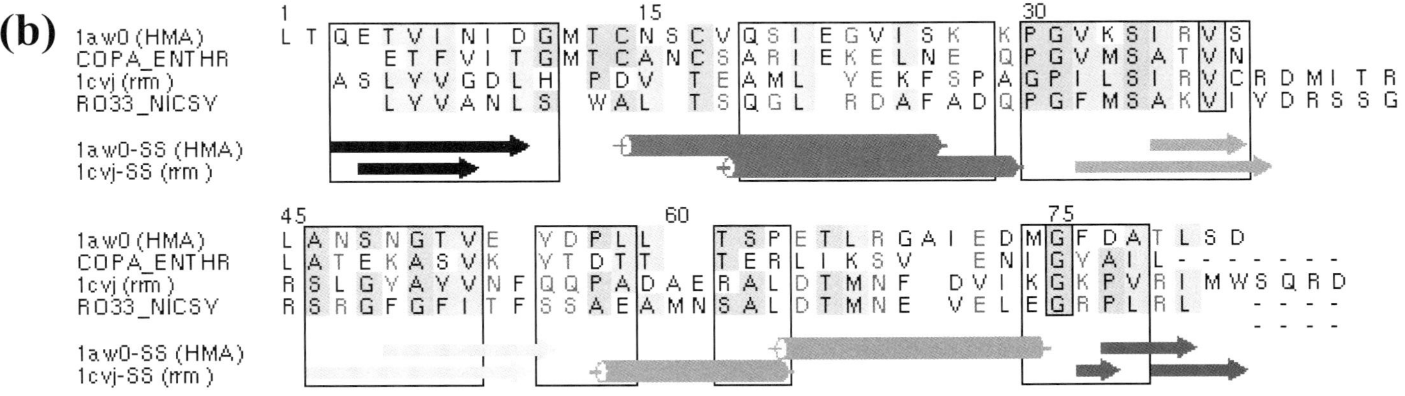

Figure 2. New potential superfamily within the ferredoxin-like fold. (a) Molscript (Kraulis, 1991) figures showing the Poly(A)-binding protein (left; 1cvj, chain F; Pfam family *rrm*) and the Copper transporter ATPase (right; 1aw0; Pfam family *HMA*) in a similar orientation. Structural equivalent regions (identified by the method of Russell and Barton, 1992) are labelled as arrows (β-strands) or ribbons (α-helices) or coil, with non-equivalent regions shown as Cα trace. Residues common to both structures are shown in ball-and-stick format. Linkage details: RMSD = 1.8 Å in 39 C_{α}–atoms; 7 identities in 12 equivalent residues; P_{3D}-value = 4.7×10^{-5} $MP_{3D} = 9.1\times10^{-4}$ (for details see Aloy *et al.*, 2002). (b) Alscript (Barton, 1993) figure showing the structural alignment of the two proteins in (a) with secondary structures (below). The best linking sequences are also shown (COPA_ENTHR and RO33_NICSY). Secondary structures are shown as arrows (β-strands) and cylinders (α-helices) below the alignment.

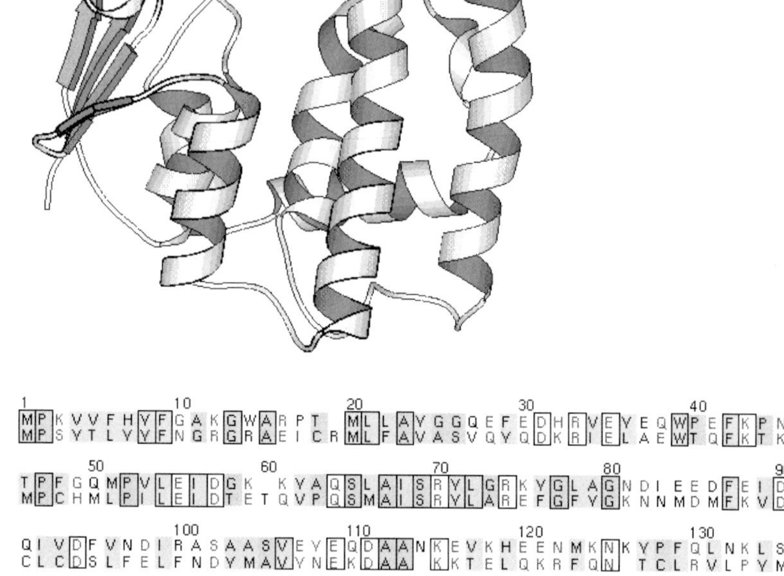

Figure 3. Sequence and structural similarity do not always mean related function. Molscript (Kraulis, 1991) and Alscript (Barton, 1993) figures showing a ribbon representation of Glutathione S-transferase (1gsq) and the sequence alignment of the former and S-crystallin SL11. Boxes show sequence identity and shading indicates similar physio-chemical properties. Although the two proteins present the same structural fold and share more than 75% sequence similarity, their functions are completely unrelated, one being a metabolic enzyme and the other a structural protein.

different functions. No previous evidence suggested a common origin for these families, but our approach linked 12 out of the 17 Pfam families assigned to this fold with significant P_{3D}-values. We were also able to infer functional features from one family to another, like the nucleotide binding site and some functional residues in the heavy-metal-associated domain (HMA) family (see Figure 2 and Aloy *et al.*, 2002 for details). This is an example of how sequence similarity after structural superposition can unveil hidden evolutionary relationships and therefore provide unknown functional details for a given protein family.

Does Structural Similarity Imply a Common Function?

Even homologous protein structures can have quite different functions (e.g. Murzin, 1993, 1996). One of the earliest examples seen is that of S-crystallin, which despite a significant sequence similarity to Glutathione-S-transferase (Figure 3), does not function as an enzyme. An extreme example is the similarity between sonic hedgehog (a signalling factor) and DD carboxypeptidase (an enzyme; Murzin, 1996).

However the problem should not be overstated. Generally proteins that have detectable similarities in sequences have a degree of functional similarity. This is less often true for structural similarities, but there are some guiding principles. Conservation of key residues in active sites usually implies similarities in molecular function, but bear in mind that there can be wide variation around a common active site theme. Proteins with similar catalytic machinery, for example, can have wildly different substrates: the proteins of the α/β hydrolase family, for example, usually contain a similar active site, but cleave substrates ranging from proteins to lipids to halogenated alkanes (e.g. chlorofluorocarbons).

One should also be aware that some folds are much more common in nature, and thus perform many different functions. This is the case for four-helix bundles, α/β barrels, or the immunoglobulin-like β-sandwich. Orengo *et al.* (1994) called these folds "superfolds", and suggested that if one observes a structure adopting one of these folds then one must exercise caution when using the observation for the prediction of function. Conversely, if one observes a protein adopting a fold that is not a superfold, it is more likely to be associated with a function, the idea being that nature has only used this fold for a specific function.

Despite showing now obvious overall functional preference, some of the "superfolds" do show a tendency to bind ligands in a common location (Russell *et al.*, 1998). α/β barrels, four-helix bundles, Rossmann or ferredoxin-like folds show preferences to bind a diversity of ligands (e.g. sugars, metals, other proteins, substrates, etc.) in a common location, while immunoglobulin-folds and β-trefoils do not.

This has been quantified at large-scale by several authors (Hegyi and Gerstein, 1999; Devos and Valencia, 2000) showing that the percentage of homologues above 25% identity performing completely different functions is 10% based on differences in the first EC digit, or 30% based on keyword description (Devos and Valencia, 2001).

We have employed a strategy that uses a structure-based prediction of protein functional sites to assess the reliability of functional inheritance (Aloy *et al.*, 2001). The problem considered is that during genome annotation, two proteins are identified as homologues, one of which (or a close homologue) has a known three-dimensional structure. In addition, the function of one of the proteins is also known and one wishes to assess whether the other protein has similar or different function. If one is able to predict a common functional site for the two proteins using the strategy proposed, the approach predicts that the two proteins will have a related function. However, if no common functional site can be predicted, the approach suggests that they will have quite different functions.

This strategy has been applied on a large benchmark giving an overall accuracy of 94%. Our results show that the three-dimensional structure is crucial for distinguishing between these difficult cases of close homologues carrying out different functions. The approach has been proved to be powerful also with very difficult cases where sequence-based methods alone have failed. A well known case is that of S-crystallin and Glutatione-S-transferase (Figure 3). These two proteins share the same structural fold and almost 80% sequence similarity (30% sequence identity). However, despite this high identity level, they perform completely unrelated functions: one is a structural protein of the eye and the other a metabolic enzyme. Our method, using structure information, was able to predict two unrelated functions for these two proteins.

This is an example of how structure-based approaches, when they are fully automated, can be linked to schemes for genome annotation significantly improving the quality of the annotation.

Comparative Modelling

The aim of comparative modelling is to use experimentally determined protein structures as templates to predict the conformation of another protein that has a similar amino acid sequence (the target). The first modelling studies, based on knowledge of homologous proteins with a common fold, were carried out in the late 1960s and early 1970s and relied upon the construction of wire or plastic models. Only later were interactive computer graphics exploited. The first published comparative modelling studies were those of Browne *et al.* (1969) on the α-lactalbumin and McLachlan and Shotton (1971) on the α-lytic protease.

Comparative modelling of protein 3D structures can now be applied with reasonable accuracy to twenty times more protein sequences than the number of experimentally determined protein structures. A protein sequence that has at least 35% identity to a known structure can be modelled automatically with an accuracy approaching that of a low resolution X-ray structure or a medium resolution NMR structure.

All current comparative-modelling methods consist of four sequential steps. The first step is to identify the proteins with known 3D structure that are related to the target sequence. This may be trivial if the sequence identity between the model sequence and the homologues of the known structure is high (more than 35%). In situations where sequence identity is low, the recognition of the fold is a very difficult task. The second step is to align them with the target sequence and to pick the known structures that will be used as templates. The third step is to build the model for the target sequence given its alignment with the template structures. In the fourth step, the model is evaluated using a variety of criteria and, if necessary, the alignment and model building is repeated until a satisfactory model is obtained.

Fold Assignment

To start the modelling process, we have to identify the template structures and define an alignment between the target and the template sequences. Any errors at this stage are usually impossible to correct later (Jaroszewski *et al.*, 2000). The structures with the highest similarity to the target sequence will be used as parents or templates. Currently, around 40% of all protein sequences could potentially have, at least, one domain modelled on a related known protein structure (Fiser *et al.*, 2000).

Originally, searches of homologous sequences were performed with local alignment programs as for example: FASTA (Pearson and Lipman, 1988); SSEARCH (Pearson, 1996) or BLAST (Altschul *et al.*, 1997) that are able to find identities shared between pairs of related sequences. Given the high rate at which new sequences become available from genomic initiatives, the importance of sensitive methods of recognising distant homologies has increased. Now, a handful of new methodologies has been implemented and successfully applied to the recognition of remote homology:

1. Threading approaches to evaluate the compatibility between the target sequence and a given structural template (Fischer and Eisenberg, 1996).

2. The use of multiple sequence alignments with a position specific scoring system, either provided by hidden Markov Models (Eddy, 1998; Bateman *et al.*, 2000), by a position specific iterative BLAST (PSI-BLAST) (Altschul *et al.*, 1997), or by searching in sequence space using intermediate sequences (ISS) (Teichmann *et al.*, 2000).
3. The incorporation of sequence profiles and knowledge-based threading potential have also improved the recognition of remote homologues (Domingues *et al.*, 1999; Jones, 1999).

Currently, PSI-BLAST is one of the most powerful tools for detecting remote evolutionary relationships. Rychlevsky *et al.*, (2000) have developed a new procedure with profile-profile searches (FFAS) that, according to the authors, gives even better results than PSI-BLAST. It is also known that any additional information, often manually introduced, about the structure, can improve the recognition (i.e., secondary structure prediction, conservation of key residues, etc).

Template Selection and Alignment

Once the fold has been assigned, one or more templates can be used depending on the degree of sequence similarity of these templates to the target protein. Usually, if the average level of sequence identity of the target to the parents is larger than 40% and the sequence spread is small between parents, then a single parent is used (Bates and Sternberg, 1998). Nowadays, the database searching methods have been tuned to find remote homologues instead of the best alignment. Therefore, although target and templates are likely to be correctly aligned if sharing more than 40% identity, they need to be realigned if they are in the "twilight zone" sharing less than 30% identity (Sánchez *et al.*, 2000). In general, PSI-BLAST correctly aligns 40% of the residues when the sequence identity is larger than 15% (Sauder *et al.*, 2000).

Once the optimal alignment of the homologous proteins has been obtained, it is then used for building a 3D model of the structure of the target. Even in the cases when the alignment is good in the protein core, the amino acids in the loop regions can be significantly displaced (Alexandrov and Luethy, 1998; Saqi *et al.*, 1999). Obviously, the key for obtaining a good level of accuracy in the modelling is to base it on alignments with few errors. A remarkable improvement in the alignment quality is obtained by using multiple sequence alignments that give additional information about which are the most conserved regions and which the most variable. Hence, the use of multiple alignment programs, such as ClustalX (Thompson *et al.*, 1999) or T-coffee (Notredame *et al.*, 2000), is of great utility. Moreover, we can profit from a number of databases of accurate multiple sequence alignments of protein domains (e.g. SMART (Schultz *et al.,* 2000), Pfam (Bateman *et al.,* 2000)) to refine our alignments and therefore to improve the quality of the final model.

Model Building

The main difference between the different comparative modelling methods is in how the 3D model is calculated from a given alignment. The original and still most widely used method is modelling by means of rigid-body assembly (Browne *et al.*, 1969; Blundell *et al.*, 1987). This method constructs the model from a few core regions and from loops, which are obtained from dissecting related structures. The assembly involves fitting the rigid bodies on the framework, which is defined as the average of the Cα atoms in the conserved regions of the fold. Another family of methods, modelling by segment matching, relies on the approximate position of conserved atoms from the templates to calculate the coordinates of other atoms (Levitt, 1992). This is achieved using a database of short segments of protein structure, energy or geometry rules, or some combination of these criteria. The third group of methods, modelling by satisfaction of spatial restraints, uses either distance geometry (Srinivasan *et al.*, 1993) or optimisation techniques (Sali and Blundell, 1993) to satisfy spatial restraints obtained from the alignment of the target sequence with homologous templates of known structure.

Modelling by satisfaction of spatial restraints is perhaps the most promising of all comparative modelling techniques because it can use many different types of information about the target sequence. For example, constrains could be provided by rules for secondary structure packing (Taylor, 1993), analyses of hydrophobicity (Aszodi and Taylor, 1996), correlated mutations (Göbel *et al.*, 1994), empirical potentials of mean force (Sippl, 1990), NMR experiments (Sutcliffe *et al.*, 1992), cross-linking experiments (Rossi *et al.*, 1995), image reconstruction in electron microscopy (Neil, 1995), site-directed mutagenesis (Boissel *et al.,* 1993), fluorescence spectroscopy, intuition, etc. In this way, a comparative model, especially in the difficult cases, could be improved by making it consistent with available experimental data.

The modelling of the structurally variable regions is still the most difficult problem in model building. It can be seen as a mini protein folding problem and it is being approached from two different perspectives: *ab initio* methods (Dudeck *et al.,* 1998; Fiser *et al.,* 2000) and database searching techniques (Oliva *et al.,* 1998; Bates and Sternberg, 1999; Deane and Blundell, 2001).

1. The *ab initio* prediction is based on a conformational search guided by a scoring or energy function (for a review see Martí-Renom *et al.,* 2000): space sampling (Moult and James, 1986), systematic conformational search (Bruccoleri and Karplus, 1987), energy minimisation (Dudeck *et al.,* 1998), molecular dynamics (Nakajima *et al.,* 2000), Monte Carlo (Rapp and Friesner, 1999), multiple copy sampling (Rosenbach and Rosenfeld, 1995), searching discrete conformations by dynamic programming (Vajda and DeLisi, 1990), self-consistent field optimization (Koehl and Delarue, 1996), and others.
2. The database approach to loop prediction consists of finding a segment of main chain that fits the two stem regions of a loop (Oliva *et al.,* 1997, 1998; Morea *et al.,* 1998). The requirements of the chosen loop of the cluster of conformations are twofold: 1) the fitting between the two bracing secondary structures, and 2) a sequence pattern presented in the target loop to model. This procedure has been successfully applied to model the structures of canonical loops of immunoglobulin complementary determining regions (CDR) (Al-Lazikani *et al.,* 1997; Chothia *et al.,* 1989; Morea *et al.,* 1998; Oliva *et al.,* 1998; Shirai *et al.,* 1999). Nevertheless, the database search is in general valid only for short and medium sized loops. Up to date classifications of long loops have failed, and it has been demonstrated that a correlation between the geometric variables describing the loop stems is needed in order to obtain such classification (Martí-Renom *et al.,* 1998).

Finally, once the backbone of the model has been built, one has to tackle the side-chain packing problem. The conformation of the side-chains are copied from a homologous template, but there is a rapid decrease in the side-chain packing conservation when the sequence identity falls under 30% (Chung and Subbiah, 1996) which implies the need of other strategies. In order to obtain a good arrangement of side-chain conformations on a fixed backbone, a combinatorial approach is often used. An important piece of information is that side-chains can be grouped in representative sets of rotamers with specific distributions. This is used as an alternative to model them. The side-chain is chosen by optimisation procedures using the mean field theory approximation (Koehl and Delarue, 1995; Jackson *et al.,* 1998). Energy-based procedures rely on the assumption that lower values necessarily correlate with more accurate positioning (Sutcliffe *et al.,* 1987a, 1987b). This puts the burden on the quality of the particular energy function used. Karplus and coworkers (Petrella *et al.,* 1998) have obtained an accuracy of around 70% on the modelling of side-chains by testing the accuracy of new force fields (Lazaridis and Karplus, 1999). They demonstrate that the absence of solvent introduces an error in the hydrogen-bonding pattern of polar residues, making necessary the inclusion of electrostatic and solvation effects. The success in the solution of the rotamer-packing problem has enabled incorporation of strategies that solve this problem in docking procedures that evaluate protein-protein interactions (Jackson *et al.,* 1998).

Model Evaluation

At present, the errors in comparative modelling are mainly due to five different sources: errors produced by incorrect template selection (Sánchez and Sali, 1997a, 1997b; Bates and Sternberg, 1999), regions without template structure (Deane and Blundell, 2001), errors due to misalignments (Saqi *et al.,* 1999), shifts of correctly aligned residues (Bates and Sternberg, 1998) and incorrect side-chain packing (Vasquez, 1996).

The evaluation of a model is critical for testing and suggesting the best and most accurate model or models. First, the model has to be checked to preserve the correct stereochemistry of a protein polymer. This can be performed with programs like PROCHECK (Laskowski *et al.,* 1998) and the clashes can be fixed by using optimisation programs based on molecular mechanics like CHARMM (Brooks *et al.,* 1983) or GROMOS (Gunsteren *et al.,* 1996). This refinement step has to be taken cautiously because the optimisation is not performed under native conditions (i.e., with no solvation, no ions and not necessarily meaningful conformation for side-chains) and it is meant simply to remove drastic and local clashes. The next step in the evaluation is the assessment of the fold (Martí-Renom *et al.,* 2000) which includes the order and length of the secondary structure elements (Aloy *et al.,* 2000; Jones, 1999; Kelley *et al.,* 2000) and the use of energetic profiles introduced by statistical criteria (Dima *et al.,* 2000; Gatchell *et al.,* 2000). In summary, these methods compare the modelled conformation to a standard structure in the PDB. Programs like VERIFY3D (Luthy *et al.,* 1992), PROSAII (Sippl, 1993), HARMONY (Topham *et al.,* 1994) or ANOLEA (Melo and Feytmans, 1998), are among those implementing this approach. Recently, comparative modelling has been applied at genomic scale (Sánchez and Sali, 1998) and domains in 58% of the 600.000 known protein sequences were modelled (Baker and Sali, 2001).

Predicting Function From Structure

Once the structure of our protein has been determined, either experimentally or by comparative modelling, the challenge is to use it to extract as much functional information as possible. In the context of structural genomics, functional information refers to both biochemical function (e.g. determination of active site residues, binding residues or specificity), and biological function that the protein performs inside the cell (e.g. subcellular localisation, interaction with other proteins, or intervention in large macromolecular complexes).

As determining the three-dimensional structure of a protein has been a difficult problem for a long time, scientists only attempted to solve it once its biochemical function was well known and proved to be important. Now, it is becoming usual that the three-dimensional structure of a protein is known before any details of its function, and computational biologists have to face the problem of predicting function from structure.

Several authors have tackled the problem of predicting important functional residues in proteins by structure-based methods (Lichtarge *et al.*, 1996; Fetrow and Skolnick, 1998; Russell, 1998; Aloy *et al.*, 2001), proving that the knowledge of the three-dimensional structure always increases the prediction accuracy over sequence-based methods. In addition, some of these structure-to-function approaches are fully automated and can be applied to large-scale analyses.

Conclusion

Although the ongoing structural genomics projects are already producing encouraging results, some regions of the structure space still remain inscrutable by large-scale high-throughput approaches. For instance, the solubilisation and crystallisation of transmembrane proteins still remain a big technical challenge. The unstructured low complexity regions, as well as flexible domain linkers, also represent a problem for the state-of-the-art techniques. All these caveats are, however, beyond the initial scope of structural genomics. In the future, much effort and technical/scientific expertise will be put into solving these problems and extending the goals of structural genomics to protein-interaction networks in order to understand the cell in all its magnitude or, at least, some large cellular organelles.

References

Alexandrov, N., and Luethy, R. 1998. Alignment algorithm for homology modeling and threading. Protein Sci. 7: 254-258.

Al-Lazikani, B., Lesk, A., and Chothia, C. 1997. Standard conformations for the canonical structures of immunoglobulins. J. Mol. Biol. 273: 927-948.

Aloy, P., Mas, J., Martí-Renom, M., Querol, E., Avilés, F., and Oliva, B. 2000. Refinement of modelled structures by knowledge based energy profiles and secondary structure prediction: Application to the Human Procarboxypeptidase A2. J. Comput-Aided Molec. Des. 14: 83-92.

Aloy P., Oliva B., Querol E., Aviles F.X., and Russell, R.B. 2002. Structural similarity to link sequence space. New potential superfamilies and implications for structural genomics. Protein Sci. 11: 1101-1116.

Aloy, P., Querol, E., Aviles, F.X., and Sternberg, M.J.E. 2001. Automated structure-based prediction of functional sites in proteins - Application to assessing the validity of inheriting protein function from homology in genome annotation and to protein docking. J. Mol. Biol. 311: 395-408.

Altschul, S., Madden, T., Schaffer, A., Zhang, J., Zhang, Z., Miller, W., and Lipman, D. 1997. Gapped BLAST and PSI-BLAST: a new generation of protein database search programs. Nucleic Acids Res. 25: 3389-3402.

Artymiuk, P.J., Poirrette, A.R., Rice, D.W., and Willett, P. 1997. A polymerase I palm in adenylyl cyclase? Nature 388: 33-34.

Aszodi, A., and Taylor, W.R. 1996. Homology modelling by distance geometry. Fold. Des. 1: 325–334.

Baker, D., and Sali, A. 2001. Protein structure prediction and structural genomics. Science 294: 93-96.

Barton, G.J. 1993. ALSCRIPT: a tool to format multiple sequence alignments. Protein Eng. 6: 37-40.

Bateman, A., Birney, E., Durbin, R., Eddy, S., Howe, K., and Sonnhammer, E. 2000. The Pfam protein family database. Nucleic Acid Res. 28: 263-266.

Bates, P., and Sternberg, M. 1998. From Sequence to Structure. In: Protein Structure Prediction: A Practical Approach. M. Sternberg, ed. Oxford Univ. Press, Oxford,UK. p. 117-141.

Bates, P.A., and Sternberg, M. 1999. Model building by comparison at CASP3: Using expert knowledge and computer automation. Proteins: Struct. Func. and Gene. Suppl. 3: 47-54.

Blundell, T.L., and Johnson, M.S. 1993. Catching a common fold. Protein Sci. 2: 877-883.

Blundell, T.L., Sibanda, B.L., Sternberg, M.J.E., and Thornton, J.M. 1987. Knowledge-based prediction of protein structures and the design of novel molecules. Nature 326: 347-352.

Boissel, J.P., Lee, W.R., Presnell, S.R., Cohen, F.E., and Bunn, H.F. 1993. Erythropoietin structure–function relationships. Mutant proteins that test a model of tertiary structure. J. Biol. Chem. 268: 15983-15993.

Brenner, S.E. 2000. Target selection for structural genomics. Nature Struct. Biol. 7 (suppl): 967-969.

Brooks, B., Bruccoleri, R., Olafson, B., States, D., Swaminathan, S., and Karplus, M. 1983. CHARMM: a program for macromolecular energy minimization and dynamics calculations. J. Comp. Chem. 4: 187-217.

Browne, W.J., North, A.C. and Phillips, D.C. 1969. A possible three-dimensional structure of bovine alpha-lactalbumin based on that of hen's egg-white lysozyme. J. Mol. Biol. 28: 65-86.

Bruccoleri, R., and Karplus, M. 1987. Prediction of the folding of short polypetide segments by uniform conformational sampling. Biopolymers. 26: 137-138.

Chothia, C. 1992. One thousand families for the molecular biologist. Nature 357: 543-544.

Chothia, C., Lesk, A.M., Tramontano, A., Levitt, M., Smith-Gill, S.J., Air, G., Sheriff, S., Padlan, E.A., Davies, D., and Tulip, W.R. 1989. Conformations of immunoglobulin hypervariable regions. Nature 342: 877-883.

Chung, S., and Subbiah, S. 1996. A structural explanation for the twilight zone of protein sequence homology. Structure. 4: 1123-1127.

Deane, C., and Blundell, T. 2001. CODA: A combined algorithm for predicting the structurally variable regions of protein models. Protein Sci. 10: 599-612.

Devos, D., and Valencia, A. 2000. Practical limits of function prediction. Proteins 41: 98-107.

Devos, D., and Valencia, A. 2001. Intrinsic errors in genome annotation. Trends Genet. 17: 429-431.

Dima, R., Banavar, J., and Maritan, A. 2000. Scoring functions in protein folding and design. Protein Sci. 9: 812-819.

Domingues, F.S., Koppensteiner, W.A., Jaritz, M., Prlic, A., Weichenberger, C., Wiederstein, M., Floeckner, H., lackner, P., and Sippl, M. 1999. Sustained performance of knwoledge-based potentials in fold recognition. Proteins: Struct. Func. and Gene. Suppl. 3: 112-120.

Dudeck, M., Ramnarayan, K., and Ponder, J. 1998. Protein structure prediction using a combination of sequence homology and global energy minimization: II. Energy functions. J. Comp. Chem. 19: 548-573.

Eddy, S. 1998. Profile hidden markov models. Bioinformatics. 14: 755-763.

Edwards, A.M., Arrowsmith, C.H., Christendat, D., Dharamsi, A., Friesen, J.D., Greenblatt, J.F. and Vedadi, M. 2000. Protein production: feeding the crystallographers and NMR spectroscopists. Nat Struct Biol. 7 (suppl): 970-972.

Fetrow, J.S., and Skolnick, J. 1998. Method for prediction of protein function from sequence using the sequence-to-structure-to-function paradigm with application to glutaredoxins/thioredoxins and T1 ribonucleases. J. Mol. Biol. 281: 949-968.

Fischer, D., and Eisenberg, D. 1996. Protein fold recognition using sequence-derived predictions. Protein Science. 5: 947-955.

Fiser, A., Do, R., and Sali, A. 2000. Modeling of loops in protein structures. Protein Sci. 9: 1753-1773.

Gatchell, D.W., Dennis, S., and Vajda, S. 2000. Discrimination of near-native protein structures from misfolded models by empirical free energy functions. Proteins: Struct. Func. and Gene. 41: 518-534.

Göbel, U., Sander, C., Schneider, R. and Valencia, A. 1994. Correlated mutations and residue contacts in proteins. Proteins 18: 309–317.

Gunsteren, W.V., Billeter, S., Eising, A., Hünenberger, P., Früger, P., Mark, A., Scott, W., and Tironi, I. 1996. Biomolecular Simulation: The GROMOS96 Manual and User Guide. Verlag der Fachvereine, Zürich.

Hegyi, H., and Gerstein, M. 1999. The relationship between protein structure and function: a comprehensive survey with application to the yeast genome. J. Mol. Biol. 288: 147-164.

Holm, L., and Sander, C. 1996a. Mapping the protein universe. Science 273: 595-603.

Holm, L., and Sander, C. 1996b. The FSSP database: fold classification based on structure-structure alignment of proteins. Nucleic Acids Res 24: 206-209.

Holm, L., and Sander, C. 1997. Decision support system for the evolutionary classification of protein structures. Ismb 5: 140-246.

Holm, L., and Sander, C. 1998. Dictionary of recurrent domains in protein structures. Proteins 33: 88-96.

Jackson, R.M., Gabb, H.A., and Sternberg, M.J.E. 1998. Rapid refinement of protein interfaces incorporating solvation: application to the docking problem. J. Mol. Biol. 276: 265-285.

Jaroszewski, L., Rychlewski, L., and Godzik, A. 2000. Improving the quality of twilight-zone alignments. Protein Sci. 9: 1487-1496.

Jones, D. 1999. GenTHREADER: an efficient and reliable protein fold recognition method for genomic sequences. J. Mol. Biol. 287: 797-815.

Kelley, L.A., MacCallum, R.M., and Sternberg, M. 2000. Enhanced genome annotation using structural profiles in the program 3D-PSSM. J. Mol. Biol. 299: 499-520.

Koehl, P., and Delarue, M. 1995. A self-consistent mean field approach to simultneous gap closure and side-chain positioning in protein homology modeling. Nat. Struct. Biol. 2: 163-170.

Kraulis, P.J. 1991. MOLSCRIPT: A program to produce both detailed and schematic plots of protein structures. J. Appl. Cryst. 24: 946-950.

Laskowski, R., MacArthur, M., and Thornton, J. 1998. Validation of Protein models derived from experiment. Curr. Opin. Struct. Biol. 5: 631-639.

Lazaridis, T., and Karplus, M. 1999. Discrimination of the native from misfolded protein models with an energy function including implicit solvation. J. Mol. Biol. 288: 477-487.

Levitt. M. 1992. Accurate modelling of protein conformation by automatic segment matching. J. Mol. Biol. 226: 507-533.

Lichtarge, O., Bourne, H.R., and Cohen, F.E. 1996. An evolutionary trace method defines binding surfaces common to protein families. J. Mol. Biol. 257:342-358.

Luthy, J.U., Bowie, D., and Eisenberg, D. 1992. Assesment of protein models with three dimensional profiles. Nature. 356: 83-85.

Martí-Renom, M., Mas, J., Aloy, P., Querol, E., Aviles, F., and Oliva, B. 1998. Statistical Analysis of the loop-geometry on a non-redundant database of proteins. J Mol. Mod. 4: 347-354.

Martí-Renom, M.A., Stuart, A., Fisher, A., Sánchez, R., Melo, F., and Sali, A. 2000. Comparative protein structure modeling of genes and genomes. Ann. Rev. Biophys. Biomolec. Struc. 29: 291-325.

McLachlan, A.D., and Shotton, D.M. 1971. Structural similarities between alpha-lytic protease of Myxobacter 495 and elastase. Nat. New. Biol. 17: 202-205.

Melo, F., and Feytmans, E. 1998. Assessing protein structures with a non local atomic interaction energy. J. Mol. Biol. 277: 1141-1152.

Montelione, G.T., Zheng, D., Huang,Y.J., Gunsalus, K.C., and Szyperski, T. 2000. Protein NMR spectroscopy in structural genomics. Nat. Struct. Biol. 7 (suppl): 982-985.

Morea, V., Tramontano, A., Rustici, M., Chothia, C., and Lesk, A. 1998. Conformations of the third hypervariable region in the VH domain of immunoglobulins. J. Mol. Biol. 275: 265-294.

Moult, J., and James, M. 1986. An algorithm for determiningthe conformation of polypeptide segments in proteins by systematic search. Proteins: Struc. Func. and Gene. 1: 156-163.

Murzin, A.G. 1993. OB(oligonucleotide/oligosaccharide binding)-fold: common structural and functional solution for non-homologous sequences. EMBO J. 12: 861-867.

Murzin, A.G. 1996. Structural classification of proteins: new superfamilies. Curr. Opin. Struct. Biol. 6:386-394.

Murzin, A.G., Brenner, S.E., Hubbard, T., and Chothia, C. 1995. SCOP: a structural classification of proteins database for the investigation of sequences and structures. J. Mol. Biol. 247: 536-540.

Nakajima, N., Higo, J., and Kidera, A. 2000. Free energy landscapes of peptides by enhanced conformational sampling. J. Mol Biol. 296: 197-216.

Neil, K.J. 1995. Structure of recombinant rat UBF by electron image analysis and homology modelling. Nucleic Acids Res. 24: 1472–1480.

Notredame, C., Higgins, D., and Heringa, J. 2000. T-Coffee: A novel method for fast and accurate multiple sequence alignment. J. Mol. Biol. 302: 205-217.

Oliva, B., Bates, P., Querol, E., Avilés, F., and Sternberg, M. 1997. An automatic Classification of the structure of protein loops. J. Mol. Biol. 266: 814-830.

Oliva, B., Bates, P., Querol, E., Avilés, F., and Sternberg, M. 1998. Automated classification of antibody complementarity determining region 3 of the heavy chain (H3) loops into canonical forms and its application to protein structure prediction. J. Mol. Biol. 1193-1210.

Orengo, C.A., Jones, D.T., and Thornton, J.M. 1994. Protein superfamilies and domain superfolds. Nature 372: 631-634.

Orengo, C.A., Michie, A.D., Jones, S., Jones, D.T., Swindells, M.B., and Thornton, J.M. 1997. CATH—a hierarchic classification of protein domain structures. Structure 5: 1093-1108.

Park, J., Teichmann, S.A., Hubbard, T. and Chothia, C. 1997. Intermediate sequences increase the detection of homology between sequences. J. Mol. Biol. 273:349-354

Pearson, W., and Lipman, D. 1988. Improved tools for biological sequence comparison. Proc. Natl. Acad. Sci. USA. 85: 2444-2448.

Pearson, W. 1996. Effective protein sequence comparison. Meth. Enzymol. 266: 227-258.

Perutz, M.F.. 1999. Glutamine repeats and neurodegenerative diseases. Brain Res. Bull. 50: 467.

Petrella, R., Lazaridis, T., and Karplus, M. 1998. Protein sidechain conformer prediction: a test of the energy function. Folding and Design. 3: 353-377.

Rapp, C., and Friesner, R. 1999. Prediction of loop geometries using a generalyzed Born model of solvation effect. Proteins: Struc. Func. and Gene. 35: 173-183.

Rosenbach, D., and Rosenfeld, R. 1995. Simultaneous modeling of multiple loops in proteins. Protein Sci. 4: 496-505.

Rossi, V., Gaboriaud, C., Lacroix, M., Ulrich, J., Fontecilla-Camps, J.C., Gagnon, J., and Arlaud, G.J. 1995. Structure of the catalytic region of human complement protease c1s: study by chemical cross-linking and three-dimensional homology modelling. Biochemistry 34: 7311–7321.

Russell, R.B., and Barton, G.J. 1992. Multiple protein sequence alignment from tertiary structure comparison: assignment of global and residue confidence levels. Proteins 14: 309-323.

Russell, R.B., Saqi, M.A., Sayle, R.A., Bates, P.A., and Sternberg, M.J.E. 1997. Recognition of analogous and homologous protein folds: analysis of sequence and structure conservation J. Mol. Biol. 269: 423-439.

Russell, R.B. 1998. Detection of protein three-dimensional side-chain patterns: new examples of convergent evolution. J. Mol. Biol. 279: 1211-1227.

Russell, R.B., Sasieni, P.D., and Sternberg, M.J.E. 1998. Supersites within superfolds. Binding site similarity in the absence of homology. J. Mol. Biol. 282: 903-918.

Rychlewski, L., Jaroszewski, L., Weizhong, L., and Godzik, A. 2000. Comparison of sequence profiles. Structural prediction with no structure information. Protein Sci. 8: 232-241.

Sali, A., and Blundell, T. 1993. Comparative protein modeling by satisfaction of spatial restraints. J. Mol. Biol. 234: 779-815.

Sánchez, R., and Sali, A. 1997a. Advances in comparative protein structure modeling. Curr. Opin. Struct. Biol. 7: 206-214.

Sánchez, R., and Sali, A. 1997b. Evaluation of comparative protein structure modeling by MODELLER-3. Proteins: Struc. Func. and Gene. Suppl 1: 50-58.

Sánchez, R., and Sali, A. 1998. Large-scale protein structure modeling of the Saccharomyces cerevisiae genome. Proc. Natl. Acad. Sci. USA. 95: 13597-13602.

Sánchez, R., Pieper, U., Melo, F., Eswar, N., Martí-Renom, M., Madhusudhan, M., Mirkovic, N., and Sali, A. 2000. Protein Structure Modeling for Structural Genomics. Nature Struct. Biol. (Suppl). November: 986-990.

Saqi, M., Russell, R., and Sternberg, M. 1999. Misleading local sequence alignment: implications for comparative modelling. Protein Eng. 11: 627-630.

Sauder, J., Arthur, J., and Dunbrack, R. 2000. Large-scale comparisson of protein sequence alignment algorithms with structure alignments. Proteins: Struc. Func. and Gene. 40: 6-22.

Schultz, J., Copley, R.R., Doerks, T., Ponting, C.P., and Bork, P. 2000. SMART: a web-based tool for the study of genetically mobile domains. Nucleic Acids Res 28: 231-234.

Shirai, H., Kidera, A., and Nakamura, H. 1999. H3-rules: identification of CDR-H3 structures in antibodies. FEBS Letters. 455: 188-197.

Sippl, M.J. 1990 Calculation of conformational ensembles from potentials of mean force. An approach to the knowledge-based prediction of local structures in globular proteins. J. Mol. Biol. 213: 859–883.

Sippl, M. 1993. Recognition of errors in three-dimensional structures of proteins. Proteins: Struc. Func. and Gene. 17: 355-362.

Srinivasan, S., March, C.J. and Sudarsanam, S. 1993. An automated method for modelling proteins on known templates using distance geometry. Protein Sci. 2: 227-289.

Sutcliffe, M., Hayes, F., and Blundell, T. 1987a. Knowledge-based modeling of homologous proteins, part II: rules for the conformations of substituted side-chains. Protein Eng. 1: 385-392.

Sutcliffe, M., Hayes, F., Carney, D., and Blundell, T. 1987b. Knowledge-based modeling of homologous proteins, part I. Three dimensional frameorks derived from the simultaneous superposition of multiple structure. Protein Eng. 1: 377-384.

Sutcliffe, M.J., Dobson, C.M., and Oswald, R.E. 1992. Solution structure of neuronal bungaro-toxin determined by two-dimensional NMR spectroscopy: calculation of tertiary structure using systematic homologous model building, dynamical simulated annealing, and restrained molecular dynamics. Biochemistry 31: 2962–2970.

Swindells, M.B. 1995. A procedure for detecting structural domains in proteins. Protein Sci. 4: 103-12.

Taylor, W.R., and Orengo, C.A. 1989. Protein structure alignment. J Mol Biol 208: 1-22.

Taylor W.R. 1993. Protein fold-refinement: building models from idealized folds using motif constraints and multiple sequence data. Protein Eng. 6: 593–604.

Teichmann, S., Chothia, C., Church, G., and Park, J. 2000. Fast assignements of protein structures to sequences using the intermediate sequence library. Bioinformatics. 16: 117-124.

Thompson, J., Higgins, D., and Gibson, T. 1994. CLUSTAL W: improving the sensitivity of progressive multiple sequence alignment through sequence weighting, position-specific gap penalties and weight matrix choice. Nucleic Acids Res. 22: 4673-4680.

Topham, C., Srinivasan, N., Thorpe, C., Overington, J., and Kalsheker, N. 1994. Comparative modeling of major house dust mite allergen der p I: structure validation using an extended environmental amino acid propensity table. Protein Eng. 7: 869-894.

Vajda, S., and DeLisi, C. 1990. Determining minimum energy conformations of polypetides by dynamic programming. Biopolymers. 29: 1755-1772.

Vasquez, M. 1996. Modeling side-chain conformation. Curr. Opin. Struct. Biol. 6: 217-221.

Wallace, A.C., Laskowski, R.A., and Thornton, J.M. 1995. LIGPLOT: a program to generate schematic diagrams of protein-ligand interactions. Protein Eng 8: 127-134.

Wootton, J.C. 1994. Non-globular domains in protein sequences: automated segmentation using complexity measures. Comput Chem. 18: 269-285.

From: *Bioinformatics and Genomes: Current Perspectives*
Edited by: Miguel A. Andrade

Chapter 10

Role of Mnemonics and Virtual Reality in Visualizing Genomics and Proteomics Data

Seán I. O'Donoghue

Abstract

Students and researchers in the life sciences face a major challenge in dealing with large volumes of rapidly growing data. To meet the challenge they require informatics systems that integrate the many diverse databases in the life sciences, and facilitate cross-querying and data retrieval. In addition, there is an increasing need for systems that automatically organize new and existing data into meaningful graphical displays that help users with understanding, remembering, and navigating through these data. Several such systems have been developed. In the future, such systems may use insights from mnemonic techniques to help users deal with the large volume of data involved. The primary method of mnemonics is firstly, to encode abstract data into concrete objects (in this case, graphical representations of proteins or genes), and secondly, to place these objects in a space with a specific and meaningful context. For proteomics data, a natural spatial context is a 'bioatlas', i.e. where proteins are located within the context of a cell, organ, or organism. While there are clear limitations to such a view, I argue that it is a good starting point. Finally, I argue that the usefulness and usability of such views will be greatly enhanced by virtual reality techniques.

Introduction

A major challenge in the life sciences today is dealing with the rapid growth in biological data. The best known examples are the sequence databases, which are increasing in size exponentially; more rapidly, in fact, than the increase in computer speed. However, many other experimental methods are being scaled up for high-throughput, e.g. protein structure determination and mass spectroscopy; thus there is an increase not just in size, but in the *number* of biological databases available each year. To deal with this rapid increase in data, we require informatics systems that can automatically integrate the many diverse databases in the life sciences, allowing the user to maintain an overview.

The most successful and best known method for integrating biological databases is to construct a 'meta'-database that integrates the main biological databases as they grow, adding cross-links between the databases; with such systems, a user can submit cross-database queries and get back formatted text summaries with links to matching entries in the original databases. Two of the best known systems of this type are Entrez (Wheeler *et al.*, 2002) and SRS (Etzold *et al.*, 1996).

However, text-based retrieval systems have their limitations. One fundamental problem is that the user needs to know in advance what keywords or commands to type, i.e. what he can get out of the system depends heavily on his prior knowledge. To formulate the problem in another way, text systems are useful, probably essential, to answer specific questions; but are not so helpful when users are unsure which questions they should be asking.

This situation is somewhat analogous to the use of computer operating systems that are text-based (e.g., MS-DOS or Unix) versus those that have graphics user interfaces (GUIs). To use MS-DOS or Unix effectively, the user needs to learn and quickly recall, say, 100 or more abstract commands (including important command options or modifiers); regular users can easily manage this task. However for many others, especially for those who use such a system infrequently, the task is somewhat daunting. Thus, in the 1980s, when it became feasible and affordable, GUI systems were developed where the user could access the same functionality by interacting with graphical

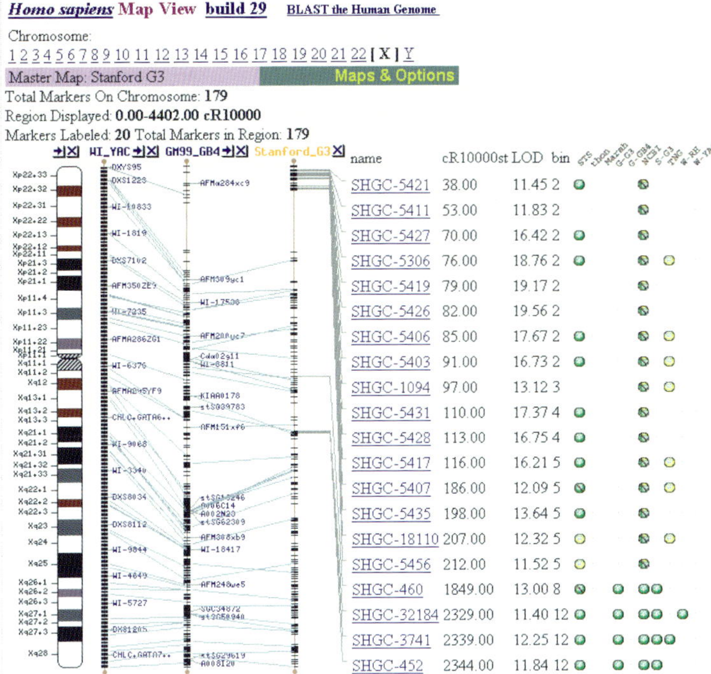

Figure 1. Graphical representation of genes on the human X chromosome. This one-dimensional representation is natural and useful, although it ignores many complexities. Screenshot from Entrez (Wheeler *et al.*, 2002).

objects; these rapidly replaced text-based operating systems as the dominate standard. Of course, each type of operating system has its strengths, and the best is a hybrid approach. However, it is fruitful to consider seriously the reasons behind the overwhelming success of GUIs versus text-based operating systems.

Two important principals behind the success of GUIs are firstly, the encoding of abstract commands or operations (e.g., 'open up a connection to a certain web server') into graphical icons, i.e. concrete forms with shapes or images that remind the user of the commands; an icon is usually easier to remember. Secondly, by arranging icons at specific places on a screen or window, the user only has to remember where the appropriate icon is located, rather than having to search his memory for the right command word. To summarize these principals: for humans, the task of understanding and remembering a large number of abstract data can be greatly aided by systems that use visual representations.

In designing an interface for bioscience databases the situation is intrinsically far worse, since we have much more data; the human genome alone has an estimated 30,000 genes, and these are thought to be spliced into

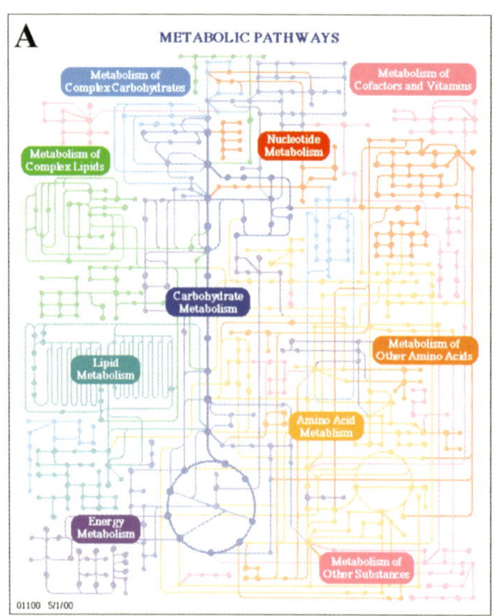
A
METABOLIC PATHWAYS
Metabolism of Complex Carbohydrates
Metabolism of Cofactors and Vitamins
Nucleotide Metabolism
Metabolism of Complex Lipids
Carbohydrate Metabolism
Metabolism of Other Amino Acids
Lipid Metabolism
Amino Acid Metablism
Energy Metabolism
Metabolism of Other Substances
01100 5/1/00

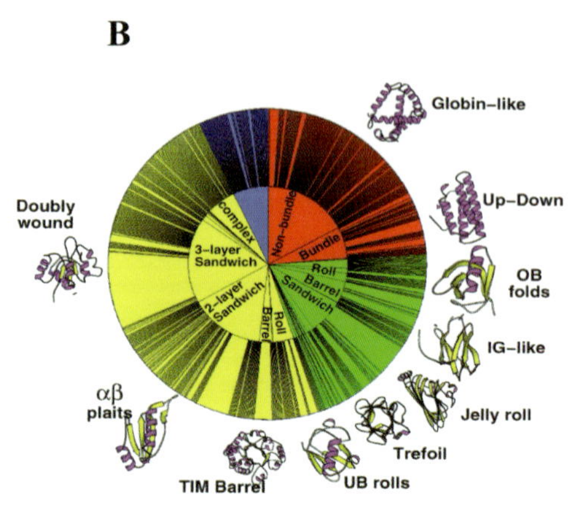
B
Globin-like
Up-Down
OB folds
IG-like
Jelly roll
Trefoil
UB rolls
TIM Barrel
αβ plaits
Doubly wound
Complex
Non-bundle
Bundle
3-layer Sandwich
2-layer Sandwich
Roll
Barrel
Sandwich

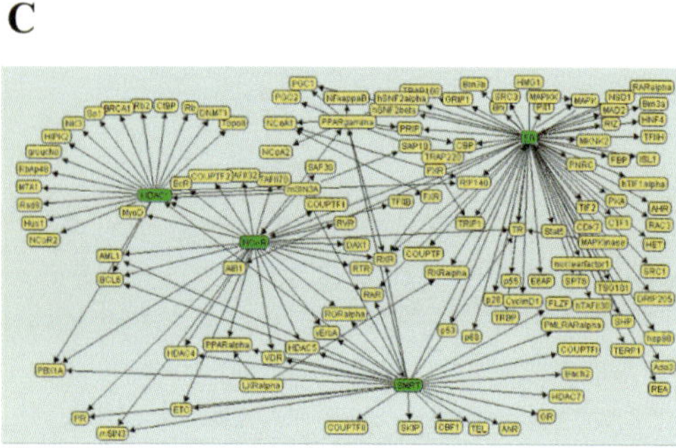
C

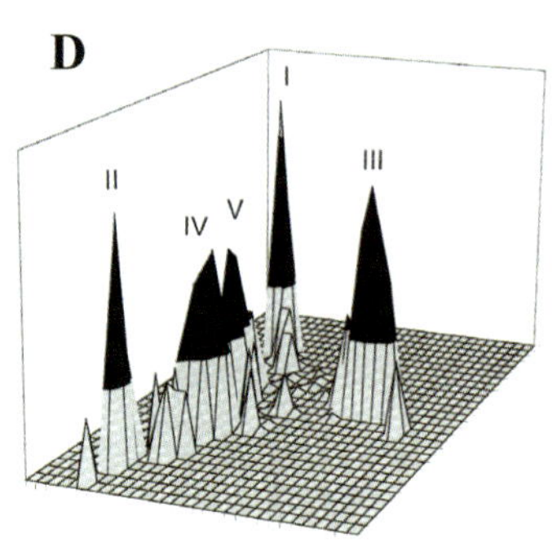
D
I
II
III
IV
V

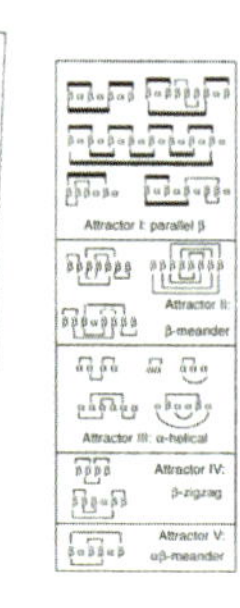
Attractor I: parallel β
Attractor II: β-meander
Attractor III: α-helical
Attractor IV: β-zigzag
Attractor V: αβ-meander

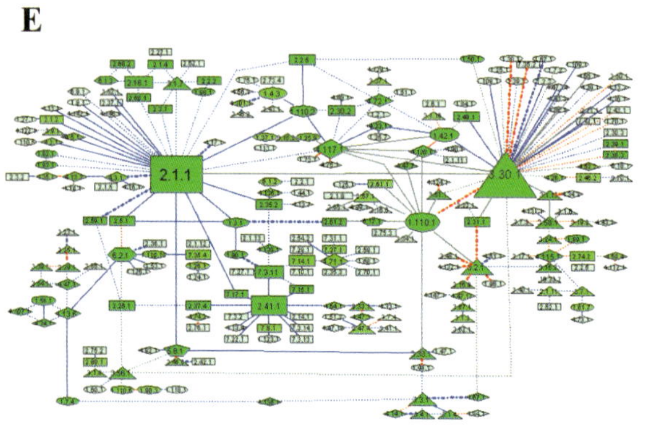
E
2.1.1

F

Figure 2. Systems of visualizing proteomic data. (A) Proteins can be visualized as enzymes and the relationship between them determined by a pathway of chemical reactions, such as recorded in the KEGG database (Kanehisa, 2000). (B) CATHerine wheel arrangement of all known protein structures into fold families (Thornton *et al.*, 1999). (C) Arrangement of proteins based on protein-protein interaction data generated by piSCOUT (www.lionbioscience.com/piSCOUT). (D) Clustering of all known protein structures based on structural similarity (Holm and Sander, 1996). (E) An arrangement of all known protein families based on known interactions (Park *et al.*, 2001). (F) One arrangement of proteins into a cellular context (O'Donoghue and Fries, 2001).

at least 60,000 proteins. This is not even beginning to count gene expression data or structural data. Even dedicated experts need help to navigate these volumes of data. In this chapter, I discuss the development of systems to define visual representations for the 'universe' of proteins and genes, and I discuss how mnemonics, and the analogy with generic computer GUIs, suggest that certain aspects are missing from the current systems. I also discuss how 3D 'virtual reality' techniques may enhance the usability and usefulness of such systems.

Defining a Genome Map

Visualizing an entire genome can be done in a straightforward way, since genes are arranged on chromosomes essentially in one dimension (e.g., see Figure 1); this is a natural and useful representation and is used in several visualization systems (e.g., Pedersen *et al.*, 2000; Wheeler *et al.*, 2002). However, this representation ignores many complexities, especially dynamic processes such as protein binding, nucleosome formation, super-coiling, and unraveling. There is clearly scope for developing more detailed and complex representations; the challenge is to do this without losing clarity.

Another important task is to visually compare several or many genomes; several such systems have been developed, essentially extending the one-dimensional representation (e.g., Glaser *et al.*, 2001).

Defining a Proteome Map

It is less obvious how to visualize proteomics data. Probably the most widely used method to represent (part of) a proteome is to arrange the proteins based on pathways (e.g., metabolic or signaling pathways; Figure 2A). There are several pathway databases, notably KEGG (Kanehisa, 2000), and also several systems for automatically assembling such data into graphical representations (e.g., Goesmann *et al.*, 2002; Karp and Paley, 1996; Salamonsen *et al.*, 1999; Minch *et al.*, 2002). In such views, proteins are usually represented either as a vertices or words. The connectivity of each protein provides a meaningful context that indicates the function of the protein.

A wealth of data is currently being produced by profiling of RNA expression. Many informatics systems are available for drawing graphs of (part of) a transcriptome based on similarities in expression profiles. These data can also be used to infer pathway interactions.

Another method of representing (part of) a proteome is by protein-protein interactions or complexes that are observed or inferred experimentally (Figure 2C). This kind of data is much easier to determine or infer experimentally than pathway data, and very recently the amount of such data has rapidly increased (e.g. Bader and Hogue, 2000; Dongre *et al.*, 2001), e.g. the recent map of protein-protein complexes in yeast (Gavin *et al.*, 2002). There are still relatively few systems available for visualizing these data (e.g., http://www.biointeraction.net, http://wilab.inha.ac.kr/protein, and http://www.lionbioscience.com/piSCOUT).

Aside from the above methods that map experimental properties, there are many other derived properties that can be used to organize sets of proteins into meaningful graphical views. For example, the 'sequence-space' view (Casari *et al.*, 1996) visualizes groups of evolutionary related proteins, with the spacing between proteins determined by sequence similarity i.e., similar proteins are placed close together.

Similar methods have been used to construct maps of the protein 'universe' (i.e. all known proteins from all organisms): the 'structure-space' (Figure 2D), where the separation between proteins is determined by similarity of 3D structure rather than sequence (e.g., Holm and Sander, 1996); PSIMAP, which arranges all known protein families by the known interaction (Figure 2E; Park *et al.*, 2001); and the CATHerine wheel (Figure 2B; Thornton *et al.*, 1999), where proteins are arranged according to their fold type.

Most of the above methods can be applied at different levels: either to visualize only a small group of proteins, a whole proteome, to compare several proteomes, or to map the protein 'universe'. The later goals, of mapping a whole proteome or all proteomes, may at first appear to be rather academic; however, even for applied research projects focused on a small number of proteins, it may often be helpful to see where these proteins fit into an overall picture.

Mnemonics: Protein Representation And Spatial Context

Currently there is no consensus proteome map that is widely in use in the field; to the contrary, the field is diverging, with new maps being proposed constantly. Here I consider some properties that future proteome maps may have. As mentioned above, a fundamental issue is the shear size of the databases. Can the human brain cope with the task of learning, say, a network of pathways for 60,000 proteins? Actually, yes: it may sound impossible, but demonstrably it is possible. Many people have a vocabulary of this many words. In fact each person of average intelligence has stored literally millions of concrete, specific details in their memories by the time they reach adulthood. The problem is not the scale of the learning task, but rather how the data is represented and organized.

Mnemonics techniques are ways of restating a learning task to make it easier. These techniques are first described in Western literature by Cicero (e.g., see Yates, 1966), however the tradition is clearly much older, being taught in the native cultures in at least Africa, the middle East, and Australia (e.g., see Chatwin, 1987), and probably most traditional oral cultures. As Western culture relied more on writing, printing, and now computers to accurately record and retrieve information, the tradition of mnemonic learning was largely dropped, although not completely. E.g., many chess grand masters use mnemonics to learn and recall tens of thousands, or even hundreds of thousands, of abstract details; closer to home, Murzin, who has reportedly memorized some 1000 protein folds, probably used mnemonics (Murzin *et al.*, 1995).

Two principals of mnemonics are: firstly, abstract details to be remembered are systematically encoded as physical objects, and secondly, these objects are mentally visualized as located at a definite position or 'locus' within a space. Applied to proteins, these principals suggest that to learn and recall a network of pathways, we should develop a systematic scheme to change each protein name or accession number into a unique physical object that can be visualized, and that at each position in the pathway, we add features that make this locus unique, so we can visualize the protein 'object' interacting with these features.

This may seem somewhat paradoxical; we actually need to remember more information than in the original data. However, human memory is anything but linear; it is actually harder for us to remember a few abstract objects like numbers, than to remember hundreds of visual images. In fact, extra visual details or actions, colors, noises, smells, etc. give us a network of redundant connections that makes learning and recall easier.

In choosing a representation and context, we are not restricted to make them biologically relevant. We are free to invent any visual metaphors, add spurious detail, distort scale, etc., provided only that it helps memory. In practice, useful systems will probably mix invented and biologically relevant information.

Defining A Bio-Atlas Map

Thus, Fries and I recently suggested that proteomics visualization systems (such as in Figure 2F) could be enhanced by using a 3D structure representation for each protein, and by modifying the layout to show the location of each protein within a cell, organ, or organism (O'Donoghue and Fries, 2001).

Protein 3D structure is a natural representation for proteins, and can be made unique with appropriate coloring and rendering. For about 40% of all sequences, a 3D structure can be predicted (e.g. see Pieper *et al.*, 2002); the rest could be displayed, e.g, as 2D objects showing predicted secondary structure. For membrane proteins, while very few structures are known experimentally, the transmembrane helices can be very reliably predicted, and hence 2D representations can be built. This would provide a visual 'icon' to represent each protein, a feature missing from the visualization systems discussed above.

We proposed using protein physical location as the principal property for embedding since it is probably the single most fundamental functional property. Moreover, this can be combined with other layouts, e.g., mapping of pathways, strengthening the spatial context information in the map compared with the other visualization methods. Cellular and tissue location are known experimentally for increasingly many proteins, and can be inferred for the rest either by homology, the presence of transmembrane helices (e.g., PHD, Rost, 1996), of signal peptides (e.g. SignalP, Nielsen *et al.*, 1997), or of characteristic amino acid composition (e.g., PreLoc, Andrade *et al.*, 1998). The simple mapping systems that we proposed are quite different from projects aimed at simulating whole cells (e.g., the E-Cell project, Tomita *et al.*, 1999). We call the collection of these maps a 'bio-atlas', in analogy to a geographical atlas, since different properties can be mapped onto the same space, and the space is inspired by the physical arrangement of proteins in the cell. Further, the method enables the user to look in detail at regions of interest (either individual protein structure or a large aggregate or cluster of proteins), while maintaining an overview of where each region fits into the context of the whole cell or organism. These maps combine protein structure and physical distribution data with other data, such as pathways or protein-protein interactions.

VIRTUAL REALITY AND 3D

Two dimensional representations can take us very far, but in 3D there is simply more 'space' in which to add detail, and more complex relationships. Virtual reality immersion systems particularly can add a depth of detail that greatly helps to differentiate one visual object from another, and one locus from another. As mentioned above, the more fine details that can be visualized for each object or locus, the more distinct they can be, and the more chance we have to remember them.

On the other hand, being land animals, we are not used to move freely in 3D; particularly, when we lose the sense of where up and down is, we lose our orientation very quickly. We are used to a kind of 'two-and-a-half'-dimensional space, i.e. two dimensions defining a ground, with a much smaller variation in the 3rd dimension. Since the point of such a system is keep orientation, we felt that such systems should best be rather 'flat' or 'squashed' in the 3rd dimension (O'Donoghue and Fries, 2001); essentially that the proteins are arranged in plane, with a 3rd dimension about several molecules 'high'. Thus the 3D effect is only really apparent when the user is somewhat 'close to' or 'zoomed into' a small number of proteins.

Several academic groups have presented generic virtual reality methods for visualizing relationships between entire databases; there have also been some efforts to customize these generic methods for biological databases (Boyle *et al.*, 1994; Ihlenfeldt, 1997). To date, however, the work published in the field have not exploited domain-specific information, such as cellular location and proteome maps discussed above.

DISCUSSION

Intuitive visualization systems that combine different types of biochemical data from a given group of proteins, say all proteins known to occur in one human cell type, would be a valuable tool both for students gaining an initial overview, and also for specialist researchers. By arranging all processes in a given cell type, or generalized cell, a specialist user can maintain an overview while choosing to drill down to specific interactions and proteins, and focus on biochemical processes. This provides a framework for keeping track of not just the particular protein that a researcher is interested in, but also of other proteins and protein systems that it interacts with. As the underlying databases grow, the researcher can see not only the results new to his field, but new results in related fields.

As with operating systems, such systems would not replace text-based query and retrieval systems, but rather complement them. Are we likely to find any truly 'canonical' organization principal for such systems, that organizes proteins in a way analogous to the periodic table of elements? I argue that a more appropriate analogy is to the geographical world atlas; when we compare proteins in a proteome, what we see is not determined by a few simple principals, but rather is the outcome of many driving forces acting over a long time. The first attempts to draw a world map where incomplete and inaccurate. However, these early exercises were probably not in vain: clearly visualizing what you believe to be true can help you to see where our map differs from reality. In geography, there is no one defining map, but different maps for different properties such as elevation, temperature, vegetation, water, etc. However these maps are so highly interrelated that one can usually be predicted from the others.

With the human genome, and many others, now sequenced, we have a map of the organization of genes within the genome, or at least well underway; the time is ripe to draft maps of the proteome. As in geography, there will be no one map, but many different properties to map. A measure of our progress will be when the different maps start to converge.

References

Andrade, M.A., O'Donoghue, S.I., and Rost, B. 1998. Adaptation of protein surfaces to subcellular location. J. Mol. Biol. 276: 517-525.

Bader, G.D., and Hogue, C.W. 2000. BIND-a data specification for storing and describing biomolecular interactions, molecular complexes and pathways. Bioinformatics 16: 465-477. http://bioinfo.mshri.on.ca

Boyle, J., Fothergill, J.E., and Gray, P.M.D. 1994. Amaze: a three dimensional graphical user interface for an object oriented database. In: 2nd International Workshop on Interfaces to Database Systems. P. Sawyer, ed. Springer- Verlag, New York. p. 117-131.

Casari, G., Sander, C., and Valencia, A. 1996. Sequence space analysis: Identification of functionally or structurally important residues in protein sequence families. In: Protein folds: a distance-based approach. H. Bohr and S. Brunak, eds. CRC Press, London. p. 124-131.

Chatwin, B. 1987. The Songlines. Penguin, New York.

Dongre, A.R., Opiteck, G., Cosand, W.L., and Hefta, S.A. 2001. Proteomics in the post-genome age. Biopolymers. 60: 206-211.

Etzold, T., Ulyanov, A., and Argos, P. 1996. SRS: information retrieval system for molecular biology data banks. In: Methods in Enzymology Vol. 266. R. F. Doolittle, ed. Academic Press, San Diego. p. 114-128. http://srs6.ebi.ac.uk/srs6/

Gavin, A.C., Bosche, M., Krause, R., Grandi, P., Marzioch, M., Bauer, A., Schultz, J., Rick, J.M., Michon, A.M., Cruciat, C.M., Remor, M., Hofert, C., Schelder, M., Brajenovic, M., Ruffner, H., Merino, A., Klein, K., Hudak, M., Dickson, D., Rudi, T., Gnau, V., Bauch, A., Bastuck, S., Huhse, B., Leutwein, C., Heurtier, M.A., Copley, R.R., Edelmann, A., Querfurth, E., Rybin, V., Drewes, G., Raida, M., Bouwmeester, T., Bork, P., Seraphin, B., Kuster, B., Neubauer, G., and Superti-Furga, G. 2002. Functional organization of the yeast proteome by systematic analysis of protein complexes. Nature 415: 141-147. http://yeast.cellzome.com

Glaser, P., Frangeul, L., Buchrieser, C., Rusniok, C., Amend, A., Baquero, F., Berche, P., Bloecker, H., Brandt, P., Chakraborty, T., Charbit, A., Chetouani, F., Couve, E., de Daruvar, A., Dehoux, P., Domann, E., Dominguez-Bernal, G., Duchaud, E., Durant, L., Dussurget, O., Entian, K.D., Fsihi, H., Portillo, F.G., Garrido, P., Gautier, L., Goebel, W., Gomez-Lopez, N., Hain, T., Hauf, J., Jackson, D., Jones, L.M., Kaerst, U., Kreft, J., Kuhn, M., Kunst, F., Kurapkat, G., Madueno, E., Maitournam, A., Vicente, J.M., Ng, E., Nedjari, H., Nordsiek, G., Novella, S., de Pablos, B., Perez-Diaz, J.C., Purcell, R., Remmel, B., Rose, M., Schlueter, T., Simoes, N., Tierrez, A., Vazquez-Boland, J.A., Voss, H., Wehland, J., and Cossart, P. 2001. Comparative genomics of Listeria species. Science 294: 849-852. http://biopath.lionbioscience.com

Goesmann, A., Haubrock, M., Meyer, F., Kalinowski, J., and Giegerich, R. 2002. PathFinder: reconstruction and dynamic visualization of metabolic pathways. Bioinformatics 18: 124-129.

Holm, L., and Sander, C. 1996. Mapping the protein universe. Science 273: 595-603.

Ihlenfeldt, W.-D. 1997. Virtual reality in chemistry. J. Mol. Model. 3: 386-402.

Kanehisa, M. 2000. Post-genome Informatics. Oxford University Press, Oxford. http://www.genome.ad.jp/kegg

Karp, P.D., and Paley, S. 1996. Integrated access to metabolic and genomic data. J. Comput. Biol. 3: 191-212.

Minch, E., de Rinaldis, M., and Weiss, S. 2002.PathSCOUT®: Exploration and analysis of biochemical pathways. Bioinformatics. In press.

Murzin, A.G., Brenner, S.E., Hubbard, T., and Chothia, C. 1995. SCOP: a structural classification of proteins database for the investigation of sequences and structures. J Mol Biol. 247: 536-40.

Nielsen, H., Engelbrecht, J., Brunak, S., and von Heijne, G. 1997. Identification of prokaryotic and eukaryotic signal peptides and prediction of their cleavage sites. Protein Eng. 10: 1-6.

O'Donoghue, S.I., and Fries, K. 2001. Method for organizing and depicting biological elements. European Patent Application EP1221671A2.

Park, J., Lappe, M., and Teichmann, S.A. 2001. Mapping protein family interactions: intramolecular and intermolecular protein family interaction repertoires in the PDB and yeast. J. Mol. Biol. 307: 929-938. http://www.biointeraction.net

Pedersen, A.G., Jensen, L.J., Brunak, S., Staerfeldt, H.H., and Ussery, D.W. 2000. A DNA structural atlas for Escherichia coli. J Mol Biol. 299: 907-30. http://www.cbs.dtu.dk/services/GenomeAtlas/index.html

Pieper, U., Eswar, N., Stuart, A.C., Ilyin, V.A., and Sali, A. 2002. MODBASE, a database of annotated comparative protein structure models. Nucleic Acids Res. 30: 255-259.

Rost, B. 1996. PHD: predicting one-dimensional protein structure by profile based neural networks. Methods in Enzymology 266: 525-539.

Salamonsen, W., Mok, K.Y., Kolatkar, P., and Subbiah, S. 1999. BioJAKE: a tool for the creation, visualization and manipulation of metabolic pathways. Pac. Symp. Biocomput. 392-400.

Thornton, J.M., Orengo, C.A., Todd, A.E., and Pearl, F.M. 1999. Protein folds, functions and evolution. J. Mol. Biol. 293: 333-342.

Tomita, M., Hashimoto, K., Takahashi, K., Shimizu, T.S., Matsuzaki, Y., Miyoshi, F., Saito, K., Tanida, S., Yugi, K., Venter, J.C., and Hutchison, C.A., 3rd. 1999. E-CELL: software environment for whole-cell simulation. Bioinformatics 15: 72-84. http://www.e-cell.org

Wheeler, D.L., Church, D.M., Lash, A.E., Leipe, D.D., Madden, T.L., Pontius, J.U., Schuler, G.D., Schriml, L.M., Tatusova, T.A., Wagner, L., and Rapp, B.A. 2002. Database resources of the National Center for Biotechnology Information: 2002 update. Nucleic Acids Res. 30: 13-16. http://www.ncbi.nlm.gov/entrez

Yates, F.A. 1966. The Art of Memory. The University of Chicago Press, Chicago.

From: *Bioinformatics and Genomes: Current Perspectives*
Edited by: Miguel A. Andrade

Chapter 11

Pseudogenes and Genomes

David Torrents, Mikita Suyama
and Peer Bork

Abstract

The knowledge on the gene content of any organism is essential for the study and understanding of its biology. The recent sequencing of large and complex genomes has forced the scientific community to develop or improve computer programs in order to identify such genes. These algorithms are based on the identification of characteristic patterns of gene-related elements (such as promoters, splice sites, polyadenilation signals, and others) and present an estimated success rate of 80%. But, neither these programs nor their evaluation procedures normally take into consideration the presence of non-functional gene copies in the genome. These dispensable gene copies, known as pseudogenes, are formed either by retrotransposition or by tandem duplication. In some cases they are difficult to differentiate by using standard procedures since they share many sequence characteristics with their corresponding functional parental genes. The only criteria used so far to identify such non-functional elements depends on the detection of either disruptions in the open reading frame or any typical sign of retrotransposition. This leads to misclassification of some genes. In order to overcome this situation, we have developed an independent strategy that is capable to differentiate many functional from non-functional sequences. This procedure takes advantage of the different selective constrains associated to pseudogenes and genes. Using this method we estimated that the human genome contains 40000 pseudogenes, doubling current approximations. We are also proposing an error rate of 23% in standard procedures of gene annotation regarding the classification of genes and pseudogenes.

Introduction

The development of sophisticated techniques that permit direct manipulation of DNA has changed the way researchers approach scientific questions related to biological processes. Around fifty years ago, classical biochemistry was restricted to the investigation of biological processes (enzymatic reactions) from a chemical point of view. Some years later, protein sequencing and purification techniques permitted the finding of relationships between these processes and particular peptide molecules. Nowadays, the capability of DNA manipulation (purification, sequencing, modification, expression in living cells, etc…) permits scientists to analyze many biological processes from a molecular point of view. The empirical generation of a large amount of information regarding DNA → Protein → Function relationships and the formulation of general biological rules offers the possibility of prediction. On the basis of this knowledge and its application to newly identified DNA sequences, bioinformaticians are able to make predictions about biological processes and relationships between macromolecules. In this sense, the recent arrival of complete genomic sequences is "happily" received by the bioinformatic community as very promising material.

Gene Prediction

In order to exploit a genome, the correct identification of the genes contained therein is required. In the case of prokaryotic genomes, the task is relatively simple. As protein-coding regions in bacterial genomes are not interrupted by intronic sequences, their identification is reduced to the detection of open reading frames by simple translation of the DNA. The key of this process is the definition of a gene in terms of the minimum accepted size, compromising the identification of the small ones. Nevertheless, this problem has been partially solved by evaluating the existing differences in nucleotide composition between coding and non-coding regions (Salzberg *et al.*, 1998).

Gene prediction from eukaryotic genomes requires much more attention and effort owing to the large size of the sequences, the low gene density (estimated to be around 1 gene/100 kb on average in the human genome; International Human Genome Sequencing Consortium, 2001) and the complexity of the gene structures (mainly the presence of introns that in higher eukaryotes can be larger than 100 kb). Alternative splicing (the possibility of expressing differents sets of exons for a given gene under different conditions), and the presence of genes nested in the intronic sequences of other genes, complicate even more the predictions. The recent release of the draft sequence of the Human genome (International Human Genome Sequencing Consortium, 2001; Venter *et al.*, 2001) and the partially sequenced mouse genome constitute a challenge for the algorithms developed to find genes in complex scenarios.

Some collections of predicted genes have been lately proposed for the human genomes (Venter *et al.*, 2001; Yeh *et al.*, 2001; Hubbard *et al.*, 2002). The methodology used in each case is quite different yielding distinct pictures of the human gene content as revealed by the little overlap observed among these collections (Hogenesch *et al.*, 2001). Usually the automatic generation of gene collections for a particular organism on the basis of its genome analysis should find a compromise between quantity and quality. If the set is too small, despite its high accuracy, it will be considered non-informative and the scientific community will not use it. On the other hand, a too large gene index is likely to contain many false positives (not true genes) and users will be skeptical about this information.

The rapid growth of the number of known cDNA sequences permits a broader identification of new genes on the basis of sequence similarity analysis. In this sense, programs that combine the identification of intron-exon boundaries with sequence similarity to known cDNAs or derived proteins, such as BLAST (Altschul *et al.*, 1997) or GENEWISE (Birney and Durbin, 1997), are highly efficient in identifying genes but too expensive in terms of time and computer power when applied to large genomic sequences. Therefore, most gene prediction strategies tend to save time by sacrificing, at least in the initial steps, the sequence similarity analysis. The programs for *ab initio* identification of eukaryotic genes (i.e. without using sequence similarity) have improved along with the available empirical knowledge on sequence patterns associated to genomic elements (for review see Guigó, 1997; Burge and Karlin, 1998; Guigó *et. al.*, 2000). In this sense, these algorithms are designed to identify protein-coding genes basically upon detection of the first and last coding exons, intron-exon boundaries, promoter sequences (mainly transcription start sites and TATA-box signals), translational signals (Kozak, 1996), and by analysing their sequence composition (using Hidden Markov Models). It has been reported that these programs, concretely GENESCAN (Burge and Karlin, 1997), present a sensitivity (proportion of true genes found) and specificity (proportion of predicted genes that are real) around 80% on average (Guigó *et al.*, 2000). It should be though mentioned that this estimation was obtained considering artificial and controlled data sets that are quite distinct from real genomic sequences, i.e. they did not contain non-functional gene copies. Since these non-functional gene copies, known as pseudogenes, normally present many of the sequence characteristics found in genes, their presence should be taken into account when evaluating the efficiency of gene prediction methodologies.

Pseudogenes Complicate Gene Prediction

We define as pseudogenes all dispensable gene copies unable to code for functional proteins (for review see Vanin, 1985; Mighell *et al.*, 2000). Those gene duplications that occur in germinal cell lines and are harmless for the organism will remain in the population and hence will be present in available genomic sequences. On the basis of the mechanism of such duplication two types of pseudogenes can be distinguished in eukaryotic genomes: processed and non-processed pseudogenes. Processed pseudogenes (also called retro-pseudogenes) are the result of a retro-transposition event in which single-stranded mRNA undergoes retro-transcription and integration in the genome with the help of the enzymatic machinery of retrotransposable elements (Esnault *et al.*, 2000). Most of these pseudogenes have lost all or part of the introns present in the parental gene and are likely to be inactive right from the time of generation since they typically present no 5'-promoter sequence. Non-processed pseudogenes are formed by partial or complete gene tandem duplication. On the one hand, fragmented gene duplicates (i.e. lacking promoter sequences or relevant exons), are likely to be born as pseudogenes, like most processed pseudogenes. Complete gene duplicates have the chance to remain active by gaining a new function (neofunctionalization), but in most of the cases they are converted into pseudogenes by the acquisition of mutations that disrupt their expression. The characteristic feature of all non-processed pseudogenes is their partial or total conservation of the gene structure of the parental gene.

How do current gene prediction strategies differentiate genes from their non-functional copies? Since pseudogenes often accumulate all possible sequence alterations with time, they may include disruptions (STOP codons or frameshifts) in the corresponding open reading frame (ORF). Standard annotation strategies catalogue as pseudogenes all identified genomic regions whenever such disruptions are detected. In addition, due to the low probability of a retrotransposed gene being correctly inserted in front an active promoter, any sign characteristic of a retrotranspositional origin (mainly the loss of introns) is also considered as indicative of non-functionality. Although the detection of these features can be used to correctly classify many real pseudogenes, we can propose some situations where these could lead to erroneous conclusions. For instance all functional processed genes (retrogenes; Brosius, 1999), all genes with pseudo-exons that are skipped during splicing, and those genes presenting sequencing errors leading to disruptions in their ORFs would be misclassified as pseudogenes. Moreover, non-processed pseudogenes with sequence alterations other than clear disruptions, e.g. amino acid substitutions or critical alterations at the promoter level, are likely to end up being annotated as genes. Although there is no quantification of such situations, as it is difficult to prove non-functionality given the infinite conditions under which a gene could be expressed, as much as 21% of current gene predictions may be affected by these cases (International Human Genome Sequencing Consortium, 2001). Therefore additional ways of discrimination between functional and non-functional sequences should be considered. On the basis of these simple criteria, pseudogenes have been identified in many organisms, from prokaryotes (Andersson and Andersson, 2001; Cole *et al.*, 2001) to higher organisms (Goncalves *et al.*, 2000; Harrison *et al.*, 2001). At present, the total number of pseudogenes is not known for any organism.

In the case of the human genome, the complete sequence and annotation of human chromosome 22 and 21 indicate a proportion of one pseudogene every 4 or 5 functional genes (Dunham *et al.*, 1999; Hattori *et al.*, 2000) suggesting a total of 7000 to 9000 pseudogenes in the complete human genome by assuming a total of 35000 human genes. On the basis of a different study where the content of processed pseudogenes was analyzed in a restricted portion of the human genome, a ratio of one processed pseudogene every 3 functional genes has been also proposed (Gonçalves *et al.*, 2000). Public databases, which contain gene predictions of the human genome, do not normally consider annotation of pseudogenic sequences as relevant. By simply searching with the keywords “human” and “pseudogene” we can retrieve up to 3534 and 767 sequences from the public DNA databases GeneBank and EMBL, respectively (by March, 2002). Since most of these entries correspond to side-products obtained from diverse genetic studies, their annotation quality is expected to be low. Despite all these surveys, there has been so far no extensive and accurate approach aiming at the identification and annotation of pseudogenes in the whole human genome. But, the number of pseudogenes, at least in mammal genomes, is likely to be high enough to demand careful consideration in gene identification procedures.

An exhaustive and accurate identification of pseudogenic sequences and an estimation of their content for available genomes are therefore needed. This information is not only relevant for an accurate identification of functional genes, but also offers the possibility to understand and quantify active genomic processes, such as gene duplication and non-functional DNA removal ("housecleaning"), which directly influence both the general evolution of the organism and the size of the genome.

Pseudogene Prediction

In order to improve discrimination between functional and pseudogenic genomic regions we have considered in a recent study (manuscript under submission) a parameter corresponding to the ratio of innocuous to deleterious nucleotide substitutions associated to a particular problem sequence. According to the neutral evolution theory, pseudogenes, like other non-functional genomic regions, are unconstrained by selection (Kimura, 1977). This means that any kind of mutation affecting a pseudogenic sequence will be harmless for the organism, possibly fixed in the population and hence detectable in available sequences. On the contrary, most deleterious mutations, i.e. negatively affecting the function of a gene, will be selected against and hardly maintained in the population. In general, the evaluation of the ratio of neutral to deleterious substitutions arising in a particular sequence involves two basic steps: (i) estimation of the number of neutral and deleterious sites at the moment of sequence formation (right after duplication), and (ii) counting the number of neutral and deleterious sites substituted thereafter. This last step must include a correction for multiple substitutions occurring at same sites.

From a practical point of view, neutral mutations are defined as substitutions that do not change the amino acid composition of the gene product (synonymous), while deleterious mutations account for those that induce amino acid replacement (non-synonymous). It is likely that some point mutations occurring in functional genes, despite inducing no amino acid replacement and therefore considered synonymous or neutral, are indeed deleterious and consequently selected against. On the other hand, substitutions inducing replacement of irrelevant amino acids and therefore considered non-synonymous are in fact fixed in the population as neutral. We believe that, although it is not possible nowadays to identify and quantify with precision neutral and deleterious sites, the approximations obtained are fair indicators of the degree of selective pressure associated to a particular sequence. These ratios, designed as d_S/d_N (or K_S/K_A), where d_S = number of

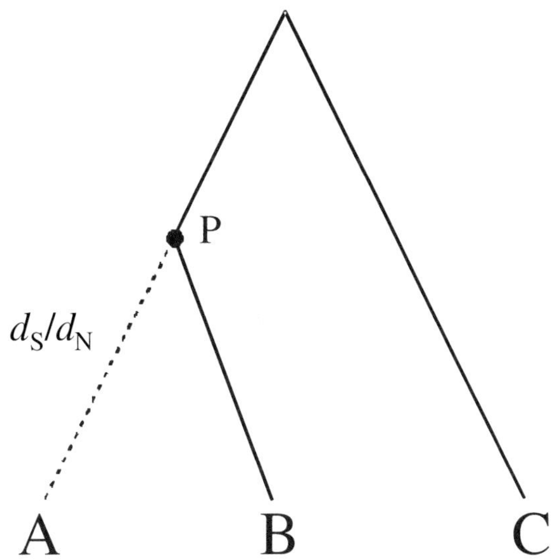

Figure 1. Assumed phylogenetic relationship between the problem sequence and the two closest functional homologues. The d_S/d_N associated to the problem sequence (A) is calculated along the dashed branch, i.e. from the parental sequence (P). P is inferred from the alignment of sequences A, B, and C using the parsimonian method (Yang, 1997).

synonymous substitutions / total number of synonymous sites, and d_N = number of non-synonymous substitutions / total number of non-synonymous sites, are expected to be about one for pseudogenes, and higher in the case of functional genes (Li *et al.*, 1981). Several analyses based on the calculation of d_S/d_N have been successfully applied to case studies to measure functional constraints associated to sequence evolution (Ohta and Ina, 1995; Nekrutenko *et al.*, 2002), but never before as a criterion to discriminate between genes and pseudogenes in genome annotation.

In this sense, we have developed a strategy to automatically obtain reliable d_S/d_N values for large data sets. Basically, this approach consists on predicting all point mutations that have been fixed in our problem DNA sequence (A in Figure 1) from the moment of its formation, i.e. from the duplication of the parental sequence (P). In order to do so, we need to deduce the nucleotide sequence of this parental gene and then to compare it with our problem sequence A. The inference of P requires two homologous sequences (B and C). The assumed phylogenetic relationship between all these sequences is shown in Figure 1. We believe that, in our study, this phylogeny accounts for the majority of the cases.

The application of similar protocols in studies of sequence evolution tends to force the comparison of the complete sequence A with very close homologues B and C. In contrast, we opted to restrict this analysis to those regions of the sequences that appear to be more conserved, i.e. regions that are expected to be under stronger selective pressure in genes. Since we can, consequently, expect higher differences of d_S/d_N values between functional and pseudogenic sequences we have permitted the use of more remote homologues B and C. In this way we increased the number of problem sequences for which it was possible to compute d_S/d_N values.

We evaluated our strategy and the reliability of the resulting d_S/d_N ratios as criterion to discriminate functional from non-functional sequences using two confident sets of functional and pseudogenic human sequences, respectively. A non-redundant (up to 50% amino acid identity) data set consisting of 3034 well annotated human cDNA sequences (the human reviewed annotation fraction of the RefSeq database, Pruitt and Maglott, 2001) was taken as the *functional set*. The collection of pseudogenes was obtained through a homology search within intergenic regions and consisted of 1730 processed elements containing at least one stop codon or frameshift in the first half of the corresponding ORFs and thus likely to be non-functional. We applied to these two sets three different methods to acquire d_S/d_N values that use different calculation models (Nei and Gojobori, 1986; Ina, 1995; Yang and Nielsen, 2000) obtaining similar results. The logarithmic distributions of these two sets according to their associated d_S/d_N values are clearly distinct as shown in Figure 2 (using the method described by Yang and Nielsen, 2000). Most d_S/d_N values associated to either functional or pseudogenic sequences are clearly indicative of stronger and weaker selective constraints, respectively. It should be noticed that positive selection, theoretically observed if $d_S/d_N < 1$, is suspected in a very few cases (Endo *et. al.*, 1996) and therefore not considered as such in this study. But, why some d_S/d_N values do not strictly follow the theoretical expectation: d_S/d_N for pseudogenes = 1, and d_S/d_N for genes > 1? Despite a certain level of inaccuracy of our method, we can think of some explanations accounting for these situations. (i) Fast evolving genes are expected to present d_S/d_N values close to one, as the number of amino acid replacement substitutions (d_N), under weaker selective constraints, can get close to the number of synonymous substitutions (d_S); and (ii) the restriction of our analysis to sequence regions with high amino acid conservation, forces d_N to remain low and therefore pushes d_S/d_N of some pseudogenes to higher values.

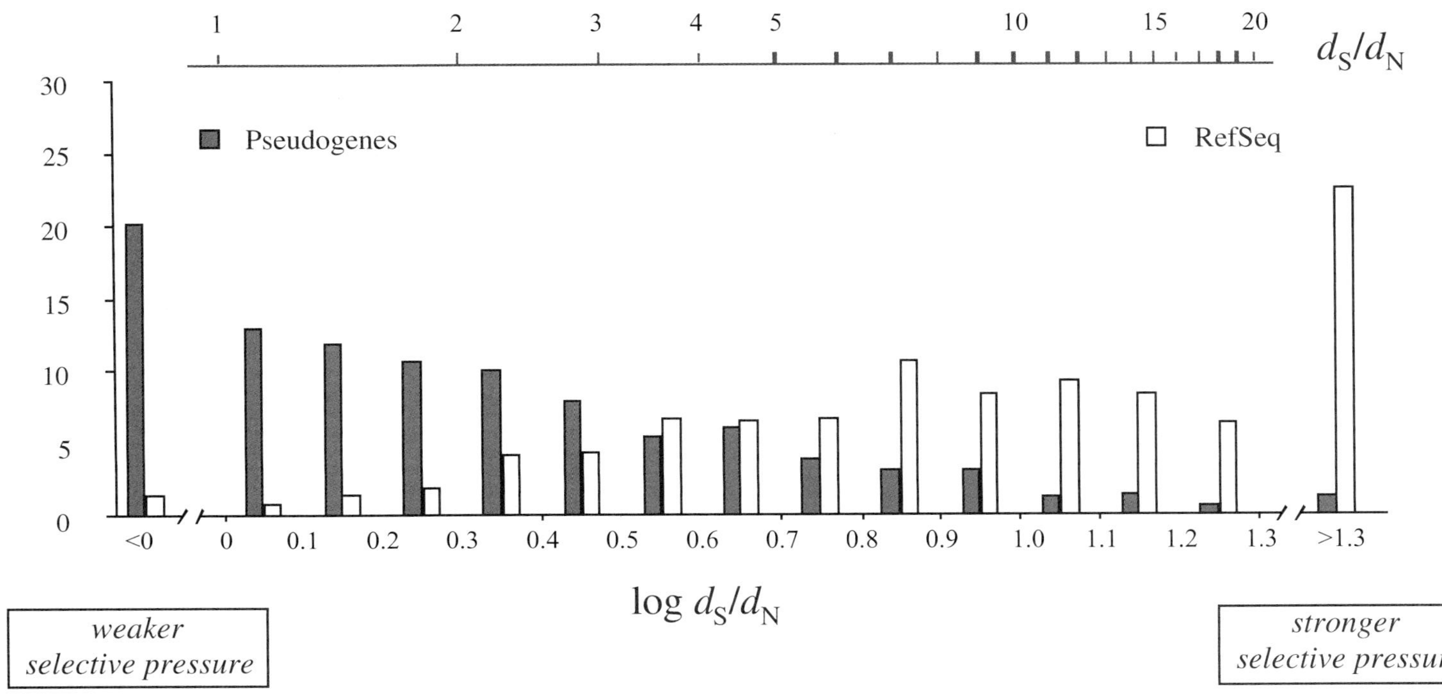

Figure 2. Distribution of reliable functional and pseudogenic sequences according to their associated d_S/d_N values.

How Many Pseudogenes Are In The Human Genome?

The information about the behaviour of d_S/d_N values obtained form functional and pseudogenic datasets can be used to estimate the portion of pseudogenes contained in any collection of human sequences with homology to known proteins. Accordingly, we carried out the same d_S/d_N analysis on a set of sequences obtained from a homology-based search through all intergenic regions in the human genome (according to ENSEMBL human gene database; Hubbard *et. al.*, 2002). Of all the sequences found, we estimated, based on the d_S/d_N values obtained, that around 10000 sequences corresponded to pseudogenes and around 2000 to functional genes. From a deeper analysis of two subsets containing either pseudogenes or genes regarding the presence of ORF disruptions with two reliable subsets of identified pseudogenes and functional sequences, our calculations indicated that up to 32% of the pseudogenes and 26% of the genes identified by standard annotation strategies (Hubbard *et. al.*, 2002) could be miscataloged as genes and pseudogenes, respectively.

We believe that this estimate of 10000 pseudogenes in the entire human genome is probably far too low. Taking into account the limitations internal to our homology search strategy (sequence similarity threshold applied and sequences lost by common DNA repeat masking), we can increase this estimate up to 35000 pseudogenes. Furthermore, we have reasons to suppose that some sequences annotated as genes in ENSEMBL database and therefore excluded from our search, could correspond to pseudogenes. Based again on the d_S/d_N analysis, we have estimated that this group of elements covers up to 23% of the whole database. We have found that most of these gene-catalogued pseudogenes correspond to non-functional partial tandem gene duplications, which have not yet acquired ORF disruptions and therefore difficult to index as such. If we add these elements, our estimate of human pseudogenes raises to 40000, covering at least a 5% of the genome.

Conclusion

We believe that 40000 is still an underestimate of the real number of pseudogenes in the human genome, since many pseudogenes are hidden to us due to their small size or degeneration under the possible level of detection by sequence similarity. The amount of non-functional DNA created by duplication seems to be higher than expected, as the high rate of pseudogene formation reflects, suggesting a relaxed evolutionary pressure on genome size in humans. This is in contrast to what has been proposed for the fly, where the high rate of non-functional DNA removal suggests a higher selective constraint on the size of its genome. However we don't know whether our genome is still growing by means of DNA duplication, or whether we have reached the "allowed" genomic size, i.e. the formation and removal of non-functional DNA are at equilibrium. A deeper analysis regarding the age of the pseudogenic regions and the rate of DNA removal can bring light to this question. Considering gene duplication as one of the important driving forces of evolution, an accurate analysis of the pseudogene content in other organism and their comparison to the human pseudogene set is needed as it can offer hints regarding the speed of genome evolution.

References

Altschul, S.F., Madden, T.L., Schaffer, A.A., Zhang, J., Zhang, Z., Miller, W., Lipman, D.J. 1997. Gapped BLAST and PSI-BLAST: a new generation of protein database search programs. Nucleic Acids Res. 25: 3389-3402.

Andersson, J.O. and Andersson, S.G. 2001. Pseudogenes, junk DNA, and the dynamics of Rickettsia genomes. Mol. Biol. Evol. 18: 829-839.

Birney, E., and Durbin, R. 1997. Dinamite: a flexible code generating language for dynamic programming methods used in sequence comparison. Proc. Int. Conf. Intell. Syst. Mol. Biol. 5: 56-64.

Brosius, J. 1999. RNAs from all categories generate retrosequences that may be exapted as novel genes or regulatory elements. Gene. 238: 115-134.

Burge, C. and Karlin, S. 1997. Prediction of complete gene structures in human genomic DNA. J. Mol. Biol. 268: 78-94.

Burge, C.B. and Karlin, S. 1998. Finding the genes in genomic DNA. Curr. Opin. Struct. Biol. 8: 346-354.

Cole, S.T., Eiglmeier, K., Parkhill, J., James, K.D., Thomson, N.R., Wheeler, P.R., Honoré, N., Garnier, T., Churcher, C., Harris, D., et al. 2001. Massive gene decay in the leprosy bacillus. Nature 409: 1007-1011.

Dunham, I., Shumizu, N., Roe, B.A., Chissoe, S., Hunt, A.R., Collins, J.E., Bruskiewich, R., Beare, D.M., Clamp, M., Smink, L.J., et al. 1999. The DNA sequence of human chromosome 22. Nature 402: 489-495.

Endo, T., Ikeo K. and Gojobori T. 1996. Large-scale search for genes on which positive selection may operate. Mol Biol Evol. 13: 685-690

Esnault, C., Maestre, J., and Heidmann, T. 2000. Human LINE retrotransposons generate processed pseudogenes. Nat. Genet. 24: 363-367.

Gonçalves, I., Duret, L., and Mouchiroud, D. 2000. Nature and structure of human genes that generate retropseudogenes. Genome Res. 10: 672-678.

Guigó, R. 1997. Computational gene identification. J. Mol. Med. 75: 389-393.

Guigó, R., Agarwal, P., Abril, J.F., Burset, M., and Fickett, J.W. 2000. An assessment of gene prediction accuracy in large DNA sequences. Genome Res. 10: 1631-1642.

Harrison, P.M., Echols, N., and Gerstein, M.B. 2001. Digging for dead genes: an analysis of the characteristics of the pseudogene population in the *Caenorhabditis elegans* genome. Nucleic Acids Res. 29: 818-830.

Hattori, M., Fujiyama, A., Taylor, T.D., Watanabe, H., Yada, T., Park, H.S., Toyoda, A., Ishii, K., Totoki, Y., Choi, D.K., et al. 2000. The DNA sequence of human chromosome 21. Nature 405: 311-319.

Hogenesch, J.B., Ching, K.A., Batalov, S., Su, A.I., Walker, J.R., Zhou, Y., Kay, S.A., Schultz, P.G., and Cooke, M. 2001. A comparison of the Celera and Ensembl predicted gene sets reveals little overlap in novel genes. Cell 106: 413-415.

Hubbard, T., Barker, D., Birney, E., Cameron, G., Chen, Y., Clark, L., Cox, T., Cuff, J., Curwen, V., Down, T., Durbin, R., *et al.* 2002. The Ensembl genome database project. Nucleic Acids Res. 30: 38-41.

Ina, Y. 1995. New methods for estimating the numbers of synonymous and nonsynonymous substitutions. J. Mol. Evol. 40: 190-226.

International Human Genome Sequencing Consortium. 2001. Initial sequencing and analysis of the human genome. Nature 409: 860-921.

Kimura, M. 1977. Preponderance of synonymous changes as evidence for the neutral theory of molecular evolution. Nature 267: 275-276.

Kozak, M. 1996. Interpreting cDNA sequences: some insights from studies on translation. Mamm. Genome 7: 563-574.

Li, W.-H., Gojobori, T., and Nei, M. 1981. Pseudogenes as a paradigm of neutral evolution. Nature 292: 237-239.

Mighell, A.J., Smith, N.R., Robinson, P.A., and Markham, A.F. 2000. Vertebrate pseudogenes. FEBS Lett. 468: 109-114.

Nei, M. and Gojobori, T. 1986. Simple methods for estimating the numbers of synonymous and nonsynonymous nucleotide substitutions. Mol. Biol. Evol. 3: 418-426.

Nekrutenko, A., Makova, K.D., and Li, W.-H. 2002. The K_A/K_S ratio test for assessing the protein-coding potential of genomic regions: an empirical and simulation study. Genome Res. 12: 198-202.

Ohta, T. and Ina, Y. 1995. Variation in synonymous substitution rates among mammalian genes and the correlation between synonymous and nonsynonymous divergences. J. Mol. Evol. 41: 717-720.

Pruitt, K.D. and Maglott, D.R. 2001. RefSeq and LocusLink: NCBI gene-centered resources. Nucleic Acids Res. 29: 137-140.

Salzberg, S.L., Delcher, A.L., Kasif, S., and White, O. 1998. Microbial gene identification using interpolated Markov models. Nucleic Acids Res. 26: 544-548.

Vanin, E.F. 1985. Processed pseudogenes: characteristics and evolution. Annu. Rev. Genet. 19: 253-272.

Venter, J.C., Adams, M.D., Myers, E.W., Li, P.W., Mural, R.J., Sutton, G.G., Smith, H.O., Yandell, M., Evans, C.A., Holt, R.A., et al. 2001. The sequence of the human genome. Science 291: 1304-1351.

Yang, Z. 1997. PAML: a program package for phylogenetic analysis by maximum likelihood. Comput. Appl. Biosci. 13: 555-556.

Yang, Z. and Nielsen, R. 2000. Estimating synonymous and nonsynonymous substitution rates under realistic evolutionary models. Mol. Biol. Evol. 17: 32-43.

Yeh, R.F., Lim, L.P., and Burge, C.B. 2001. Computational inference of homologous gene structures in the human genome. Genome Res. 11: 803-816.

From: *Bioinformatics and Genomes: Current Perspectives*
Edited by: Miguel A. Andrade

Chapter 12

Bioinformatics and Genomes: The Future

Contributions From All Authors

Introduction

Predicting how the emergence of a new research area may affect biology or even the entire world is not really what we may call an exact science. Actually, the vast majority of such predictions made in the course of the last century proved to be quite short sighted. However, it is important to try! Without looking too much ahead, all the authors in this book have a clear intuition of what seems to be the genuine driving force in their respective fields of speciality. Albeit it may be difficult to tell the exact destination one may at least point out the directions these forces seem to be aiming at.

Genome Analysis: What For?

The driving force of genome analysis is extremely new and still loaded with an incredible amount of enthusiasm and energy. This has resulted in a high level of inflation on several accounts: increasing generation of basic data yielding the complete sequences of more and more organisms, and an explosion of the number of technical and scientific groups (academic and commercial) devoted to the collection, analysis, and interpretation of this new data. The perceived academic and commercial value of new genomes is also increasing fast, as can be judged by the vast amount of new projects that rely on genome analysis and comparative genomics.

Sceptics watch this process with contempt and suggest that the analysis of genomes was given much more importance than it deserves. "Generating so much data of which no one can make any sense has been a terrible waste of resources" is their usual claim. And it is true that one cannot deny a few facts: the ease with which data has been collected has taken most of those working in the field of bioinformatics by surprise. In the 1990s, the most extreme optimists among us did not expect the coming of the human genome before 2008, and so there would be plenty of time to think of new means of analysing the data. However, the effectiveness of scientific political lobbying was overlooked. Politicians managed to free vast amounts of resources, paving the way for the technological improvements that made everything possible. There is no denying that this success owes a lot to another simple fact: genomes are sexy and their importance is easy to sell to non-specialists. Unfortunately, finding the resources to develop the algorithms needed for further analysis proved much more of a headache. Sequence analysis algorithms are not so sexy when it comes to politics; and they rarely make it to the tabloid headlines.

All things considered, it is therefore not so surprising to see that the development of methods has taken up a much slower pace than the generation of data. Is it only a matter of time until methods catch up? Jens Kleinjung (NIMR, London) agrees with the opinion that we are now mostly in the learning phase, and thinks that the impact on applications is still mostly wishful thinking. Cédric Notredame (IGS, Marseille) points out that biologists have always been extremely creative when it comes to replacing theoretical analysis with data collection: for instance EST sequences proved to be a reasonable substitute to ineffective gene predictions just as massive two-hybrid systems may prove another practical shortcut toward the understanding of protein-protein interaction. The irony, of course, is that each of these 'shortcuts' generates even more data.

In any case, we think the current efforts in generating data and developing methods to analyse the data are justified by promising applications. Eric Minch (LION bioscience AG, Heidelberg) foresees three major fields of practical application: plant crop protection and yield enhancement, attacking pathogenic organisms, and elucidating the mechanisms of disease states and therapeutic agents. Patrick Aloy (EMBL, Heidelberg) notes other applications from genomic analysis: the importance of gathering as many sequences as possible in order to cover the sequence space and prioritise the targets that have to be tackled by structural genomics projects. The coverage of the sequence and structure space will indubitably provide important insights to the understanding of protein function and evolution. Sean O'Donoghue (LION bioscience AG, Heidelberg) indicates that bioinformatics is also needed not only for the analysis of the data, but also to aid the analysis itself by providing tools for the visualisation of genome information. Other more long-term benefits of genome research are envisaged by Eric Minch such as answering scientific questions like "How do organisms work?" or "How are they related phylogenetically?"

Interaction With The Wet-Biologist

But this is our view. What about that of the general community of biologists? How do they perceive the possibilities from the genomic projects, and from the computational analysis of these genomic data? In this respect, Frank Eisenhaber (IMP, Wien) notes that there have been several booms in biological subjects (enzyme kinetics, population dynamics, protein folding studies, etc.) that promised to flourish into some sort of theory but they all ended up in a disappointing way. It turned out that such new knowledge was generally not used for making breakthroughs in molecular biology. Will the same happen with genomes?

We do not think so. Enrique Morett (UNAM, Cuernavaca) indicates that the accumulation of data on gene order allows predictions of operons that can be experimentally tested. Jens Kleinjung thinks that the introduction of the concept of (macromolecular) sequence as the expression of a language is profoundly changing biology, putting a soft science closer to the harder sciences like physics and chemistry. Eric Minch shares the feeling that molecular biologists seem to be developing a more sophisticated and general view of their field, and that they now widely (though still not universally) accept the possibility of advancing molecular biology through bioinformatics. Christian Reich (Millennium Pharmaceuticals Inc, Boston) remarks here that biologists are aware that the days of the "one gene - one PhD student - one paper" kind of approach are gone.

And then, ironically, the reaction of the biologists to the "invasion" of bioinformatics into the realm of the biological interpretation of data is twofold, as Eric Minch notes: on the one hand, there is relief at the solution of problems of handling and analysing massive amounts of data; on the other hand, there may be a certain pessimism that computational techniques can dominate or even replace the intuition of an experienced biologist.

And the acceptance of bioinformatics by the whole community of biologists is crucial, as Carolina Perez-Iratxeta (EMBL, Heidelberg) points out, since the main goal of the computational analysis of genomes is the acceleration of the biological research. In this sense, she thinks that non-experts in bioinformatics are too afraid of the apparent complexity of the genome analyses. They should be able to put a bit of an effort into using computational tools and services instead of relying on bioinformaticians. In this respect, David Torrents (EMBL, Heidelberg) points out the importance of developing the teaching of bioinformatics and related disciplines like programming and statistics. Cédric Notredame agrees with this vision and states that biologists should not anymore consider themselves as chemists doing biology, they must also become statisticians doing biology.

Patrick Aloy predicts a reversal in the relation of dependence between computational analysis and experiment. In these days, the bioinformatics analysis of genomes is seen as a service for the experimentalist. He foresees the development of hybrid computational-experimental groups where a tight collaboration between wet- and dry-biologists leads to quick experimental analysis of computational predictions.

Coping With Data (Or Not)?

Miguel Andrade (EMBL, Heidelberg) thinks that the resistance of the general biology community to accept genomic data is not just a problem of inertia to adopt something new. There is also a reasonable distrust since the quick development of the techniques applied to the analysis of genomes has produced some misinterpretations, or a lack of them.

David Torrents (EMBL, Heidelberg) notes in this regard the sequencing of the human genome. Although the genome sequence is apparently complete the prediction of human genes is grossly imperfect, and he is afraid that the re-analysis of the genome will last at least for a decade. Regarding the latter point, the sceptic could add that there have been few outstanding successes so far, deriving from the completion of the human genome (or from other eukaryotic organisms like the worm *Caenorhabditis elegans* or the fly *Drosophila melanogaster*). Eric Minch justifies this by indicating that already most of the simple systems of either sort have been discovered and exploited before large-scale genome sequencing began, and the theoretical tools for exploiting the genomic data relevant to the remaining complex systems have not been widely applied or in most cases even developed.

Christian Reich thinks that some of the deficiencies in bioinformatics techniques applied to genome data are due to the lack of integrity, completeness and quality of the genome data itself. He requests a high-quality, well annotated and coherent set of genes, with connections to other, non-gene-centric information such as biological pathways, systems, diseases, and phenotypes.

The effect, as Miguel Andrade notes, is that there is a data jam in the road that starts from the raw genome sequence (easy to derive), through the gene prediction (where the reliability of the data drops), through protein function and structure prediction (where the experimentally proven predictions are scarce), to the expression levels of the genome, and metabolic pathways (with scarce and heterogeneous data regarding only some organisms).

Accordingly, Eric Minch notes that unlike the situation with genome analysis, where there is a flood of reliable data, the situation is very different for the genomic expression analysis. He believes that despite great advances in this area, this type of analysis remains bedeviled by problems of low resolution, large errors, and difficulties in replication, comparison, and normalization. In this respect, Javier Tamames (Alma Bioinformatica, Madrid) remarks that stable and homogeneous databases, such as those available for macromolecular structures, may still be years away.

In the field of analysis of regulatory networks, Eric thinks that the problem is again the lack of data since, although several promising formal techniques for reconstruction of gene regulatory networks from expression data have been developed, there are not enough data available at high sampling frequencies for these to be applied with much hope of accurate predictions. The existence of hidden variables (protein and metabolite concentrations, or protein "states") permits multiple explanatory models. The problems of data quality and paucity also affect the related, non-genomic data of proteomics and "metabolomics", and the analysis of polymorphisms and their relationship to phenotypes.

Academia And Industry: A delicate Co-existence

As a reaction to the pressure of data, the bioinformatics community has evolved through introspection (development of better methods) and expansion (including the starting of new companies). Miguel Andrade, as an academic (but with many former colleagues in companies) watches this process with amazement and wonders whether the coexistence between academics and companies competing for the same data, but for different resources, will lead to symbiosis, parasitism, or phagocytosis. Carolina Pérez-Iratxeta expresses her concerns about the proliferation of companies inside academic environments.

Eric Minch sees that the obvious approach is for industry to outsource to academia, a model that is historically and currently plagued by the conflict between academic freedom and industry's desire for proprietary intellectual property. He thinks that there is probably no general solution for this conflict of interests, but rather only case-by-case successes and failures.

Eric Minch explains pragmatically the basic difference of mandates between these two communities: companies exist to create products and services, which they can sell at a profit; academic institutions exist to pass along and extend knowledge. Successful companies have proprietary knowledge; academia succeeds through knowledge sharing. What companies can therefore offer is a set of problems whose solution will generate some income; what academics can offer is access to intellectual resources for the solution of problems.

Christian Reich makes these points particular to bioinformatics. He sees the superiority of companies over academia through their access to high throughput technologies that allows them to validate sequences, transcripts, and expression profiles at high turn-over rates. But he also remarks that these data needs to be made available publicly. Academia is much better in tackling basic science problems, among them bioinformatics analysis algorithms. It also needs to continuously produce biological facts and findings, which in turn need to be better organized and structured to be helpful in analysis. Jens Kleinjung joins to praise the work in academia for its abundance of freedom and time to pursue ideas that may or may not work. Cédric Notredame points out that the rising cost of equipment and infrastructure is putting a tremendous pressure on academia. This causes many projects to become complex political and financial assemblies, forcing them, via the financing bodies, to be accountable to the industrial world, which he thinks does not provide a proper environment for the production of innovative ideas.

Concluding Points

Data is being derived from genomes and it will be continue to be generated for a while no matter what the applications are. This data, says Cedric Notredame, will necessarily lead to interesting results sooner or later, just as the accumulation of protein sequences in databases leads to the delineation of protein families and to the derivation of remote homologies by using profiles on multiple aligned sequences. Many of the objects we deal with in biology are only defined by comparison to other objects (such as sequence domains). There is no doubt that each new genome will challenge (at least partially) the interpretation we make of all the available genomes, up to a point... but we are not there yet.

Eric Minch observes that although algorithms for biological sequence comparison and structural prediction have been relatively successful in their application to homology-based functional prediction, they have reached their intrinsic limits. The hypotheses proposed by computational analysis are just that, hypotheses that have to be tested by experiment. At least, the computational analysis can point to the most plausible or interesting hypotheses. Necessarily, bench scientists and bioinformaticians will have to work in a closer collaborative effort in order to investigate the properties of biological systems and their functions; exploiting the strengths of both groups.

Synthesising our most optimistic thoughts, Jaap Heringa (NIMR, London) notes that enormous potential lies ahead when all of these methods and model systems have been developed. In addition massive and, through converging technology, relatively consistent sources of data are going to be integrated. Together these will allow humanity for the first time to gain a deeper understanding of the regulatory complexities within organisms that live and evolve through time as a result of their genome and environment. Such integrative endeavours will also allow the evaluation of the bioinformatics tools and subsystems at this higher level, which is expected to lead to new ideas and inspiration concerning the basic tools and further building blocks.

The data is being generated. Happy or not with it, we have to use it. Further developments in methodology are needed as well as a better understanding from the general biologist of the real possibilities given by the analysis of genomes. This will probably require the coordinated efforts of bioinformaticians and experimentalists, both from academia and industry. A failure to interpret the data may result in an end to genome sequencing. We cannot allow this to happen because we may loose a unique opportunity to improve the understanding of biology in a focused way.

Index

D

E

F

G

H

I

K

L

M

N

O

P

Q

R

S

T

U

V

W

X